Sampling Techniques
An Applied Approach

NIPA® GENX ELECTRONIC RESOURCES & SOLUTIONS P. LTD.
New Delhi-110 034

About the Authors

Ajit Sharma is working as Assistant Professor (Statistics) at Dr Yashwant Singh Parmar University of Horticulture and Forestry, Nauni, Solan, Himachal Pradesh having 5 years of experience in teaching, research and extension activities. He has attended various national and international seminars, conferences, symposiums and various trainings in statistical tools used in agricultural science and has ample experience in handling statistical software like IBM SPSS, STATA, R, Design Expert, STAR and Python etc. He is presently associated with Patanjali Research Foundation-Haridwar (UK-India) as an expert for their analytical work projects. He has published about 25 research papers in various national and international journals in the field of Statistics and Economics. His present area of research includes econometrics, regression analysis, time-series analysis, design of experiments, sampling techniques, machine learning and multivariate analysis etc.

Shilpa is an experienced Assistant Professor working in the department of Social Sciences at Dr Yashwant Singh Parmar University of Horticulture and Forestry, Nauni, Solan, Himachal Pradesh for over six years. She is the former gold medalist in her field in the same university at her master's level. Apart from her skills in teaching, research, SPSS, STATA, E-Views, Lingo and other analytical tools of data analysis, she was awarded many times by various organizations such as Himotkarsh and SADHNA Society for her excellence in academics and research work. She currently teaches many posts graduate courses including Advanced Econometrics and Production Economics in the University and mentoring five postgraduate students. Shilpa's writings covered a variety of areas including, marketing and trade, econometrics and regression analysis. She is currently handling an international project titled" "Apple production under climate change: Willingness to pay for weather indexed insurance" sponsored by SANDEE-ICIMOD, Nepal.

Sampling Techniques
An Applied Approach

(Highly useful for Bachelors, Masters and Ph.D. Scholars)

Ajit Sharma
Assistant Professor (Statistics)
Dr. Yashwant Singh Parmar University
of Horticulture and Forestry, Nauni
Solan, Himachal Pradesh, India

Shilpa
Assistant Professor (Agricultural Economics)
Dr. Yashwant Singh Parmar University
of Horticulture and Forestry, Nauni
Solan, Himachal Pradesh, India

NIPA® GENX ELECTRONIC RESOURCES & SOLUTIONS P. LTD.
New Delhi-110 034

NIPA® GENX ELECTRONIC RESOURCES & SOLUTIONS P. LTD.

101,103, Vikas Surya Plaza, CU Block
L.S.C.Market, Pitam Pura, New Delhi-110 034
Ph : +91 11 27341616, 27341717, 27341718
E-mail:newindiapublishingagency@gmail.com
www: www.nipabooks.com

For customer assistance, please contact
Phone: + 91-11-27 34 17 17
Fax: + 91-11- 27 34 16 16
E-Mail: feedbacks@nipabooks.com

ISBN: 978-93-94490-52-9

Composed and Designed by NIPA®.

Preface

Sampling provides a comprehensive overview of the research, including the basics of sampling theory and practice, as well as research-based topics and examples of emerging trends. The text is useful for basic and advanced research sample studies. Many other books available to graduate students do not contain the latest developments in the research sample area. Many of the topics discussed here are not found in other books. In each section, ideas are illustrated with numerical examples. Sampling includes the basics of sample theory and practice, as well as research-based topics. The method by which the researcher selects a sample is known as the sampling method. It is important to understand the different sampling methods used in agricultural studies. Sample techniques are part of research that plays a major role in validating research results. Without a good sample, good research practice is not possible. Research is a systematic effort aimed at solving problems that play a major role in the development of a country and without good research the development of a country is impossible. The designed book provides an easy understanding of sampling techniques used in applied fields such as agriculture, horticulture and forestry etc. This book covers a wide range of topics depending on their functionality, mathematical ideas and practical examples. Some of the important topics discussed here are not found in other books. Ideas are illustrated with appropriate theoretical examples and numbers with continuous accuracy.

This book will be useful for undergraduate, postgraduate students and also for researchers in the field of mathematics, statistics / agricultural statistics, economics / agricultural economics and other applied fields. This book will help for the desire to learn the basics that can be powerful without difficulty in the field of sampling analysis. It is the most important instruction for official statistics and statistical calculations used for non-response and small area limitations are now considered to be a part of the sampling methods. As a result, partial coverage between the sample and other statistical areas has increased, and reliance on the other makes it important for the sample to be a basic mathematical component. Sampling methods are important in the construction of surveys and tests, to validate results, thereby studying mathematics, social science, and various other fields that use sample data. However, most of the literature available on this subject is very advanced and informative or widely used, descriptive, and has no statistical implications. Sampling methods and applications provide a balanced, effective treatment of the techniques and application of the most widely used sampling procedures for reduced populations. It keeps statistics down to a minimum, but does not completely avoid them: it incorporates the results of the terms within the text but provides them based on each chapter. According to an easy-to-follow, step-by-step presentation, the author promotes each topic with illustrations followed by examples and exercises. All of this is built from everyday, practical situations covering a wide range of topics, from education appraisal to health care costs and presidential elections. Why should we go for the most advanced, theoretical houses where our needs are straightforward, practical knowledge? Why do we risk losing important mathematical results that are often left out in basic texts? This book has everything we need, presented clearly and logically to quickly access the topics that are at the heart of reality. Lastly, we would like to thankfully acknowledge the NIPA, my parents, teachers and friends.

Authors

Contents

1

Introduction

Basic thought of Sampling

The data is very important in every field for decision making whether it is medical sciences, engineering sciences, agricultural sciences and management sciences etc. for this we need to plan the correct data. If we don't have correct data then based on that we cannot plan good decisions and it will provide misleading results. So, to achieve or plan the right data it is necessary to have the right technique which is to be used while collecting data and an appropriate method of data collection is preferred. From this, the two important questions arise (i) what are those methods required to obtain paramount data or sample and (ii) how best to use the data or sample to estimate the characteristic of the whole population i.e., what are those methods by which we can study our population parameters to draw a proper statistical inference.

For any sampling strategy there are two important steps (i) selection procedure i.e., how the data or the sample is obtained from the population and (ii) estimation procedure i.e., the formulas or the methodology to be used for the estimation of population characteristics by using the sample information.

In India, the census survey is done after every ten years, but in between these years what will a researcher / a data analyst / a scientist do? Can he or she may wait for the next ten years, the answer is 'NO'. In that situation, the basic thought of sampling arises because to study the whole population we can concentrate on only a part or a fraction of that population. For this fraction study, we can apply various methods of sampling which we are discussing in this book. Sampling enables a lot of research either in agricultural or other allied fields. It is one of the vital factors which decide the accuracy of our research/survey result. If something is going wrong along with our sample then, it will likely be immediately contemplated in the final result. There are a lot of techniques which help us to acquire the sample depending upon the need and situation. This book explains some of those techniques. A statistical investigator is often concerned with studying some characteristics of a population because complete enumeration cannot be observed for some practical reasons. In our agricultural sciences / applied sciences we need to collect data to answer their research problems said, so where these sampling techniques are supposed to be used and what kind of data do we need to collect from the survey. Mainly we collect three types of data i.e., survey data, experimental data and observational data. To collect this survey data, we use sampling techniques.

Survey data: It is that type of data where we require some information but this information is based on unknown data to the researcher but known to the respondents.

Example: Income earned by the farmer from various farming activities is typically collected from a representative sample chosen.

Experimental data: In this, the information sought is based on data that requires to be generated.

Example: Performance of a new variety and collected from carefully controlled experiments.

Observational data: Observing natural phenomena with or without interventions.

Example: Air pollution level and collected directly.

In this book, we mainly focus on various sampling techniques generally applicable to survey data and this data is collected through sample surveys.

Survey: A survey is a systematic method of gathering information from units or samples to provide descriptive/inferential measures for the entire group or population to which these units belong to. By obtaining this sample information we can get an idea about the entire population.

For example: In this pandemic called COVID-19, we see that at least two or three surveys were conducted to assess the antibodies percentage in people from different locations. This is done because we cannot go for the entire population to enquire whether the antibodies are present or not in every individual. Then to find out the proportion in our sample that how many are infected or have the antigen and then use this collected information from the sample somehow, we need to predict the entire population and come to a conclusion whether this pandemic affects the city or that particular area as a whole.

Various Examples

- If the statistician or researcher wants to evaluate the per capita income of Himachal Pradesh, he or she will have to enumerate all the earnings of the individuals of the state, which is rather a very difficult task.

- To estimate the average height of adult males in district Solan, the investigator cannot afford the time or expenses required to measure the height of each adult male in Solan.
- A farmer wants to know what proportion of his tomato crop is diseased, and cannot afford the time and expenses to examine every tomato.
- If a researcher wants to know the impact of the New Education Policy in Himachal Pradesh, it is not possible or feasible to collect information from each student in every school.
- In forest surveys, it is quite tedious for a researcher to enumerate every tree.

In all such cases, the investigator should select a limited number of individuals by applying some sampling technique or statistical procedure to collect the information from them. Thus, sampling theory consists of selecting a proportion of the population in the sense that the sample will reveal the main characteristics of the population.

Population or Target population: The population is the entire group of items or individuals about which information is collected. Its size is denoted by N.

Various Examples

- Collection of households in an area, where the households are characterized by their addresses and names of the heads of households etc.
- Collection of industrial units in urban areas.
- In a political poll, the target population will be all adults, whose names are registered in the voter list.

Note: A population can be finite (countable) or infinite (uncountable) depending on whether it contains a finite or infinite number of units.

Examples of Finite Population

- Number of students in class.
- Number of employees in an office.

Examples of Infinite Population

- Stars in the sky.
- Hairs in the human head.

Method of obtaining information about population

The required information about the population can be obtained in two ways:

1. Census Survey or Complete Enumeration
2. Sample Survey

Census or Complete Enumeration

In a census survey, the required information is collected from the whole group of population. Thus, the procedure to observe the entire population is called census or complete enumeration. In this, each and every unit from the population is to be studied, recorded and analyzed i.e., Census of a country, Census of import and export etc.

Census investigation is useful in the following situations when:

- The study area is limited.
- A deep study is to be performed.
- An adequate level of accuracy and reliability is required.
- The investigator or researcher has resources.

Sample Survey

Most of the time, it is not possible to test everything in the target population, and sometimes it is possible to get good enough results by reading only a fraction of the population. In such cases, a sample survey has a great utility over censuses. Thus, the procedure to observe some part of the population is called a sample survey. Conducting a census survey is impossible when the population is infinite. A high-quality sample survey can produce better estimates with lesser errors.

Objective of Sample Survey

The objective of the sample survey is to make inferences about the population parameter from the information contained in the sample. Two factors affect the quality and quantity of information contained in the sample.

1. Representativeness of the sample
2. Appropriate sample size

Sample: Sample is a fraction / compartment / subset of the population. It should be represented by the whole population understudy to have an idea or to take some decision about the population. Its size is denoted by n.

Example: In an opinion poll, a relatively small number of persons are interviewed. Their opinions on some current issues are collected to get an idea about the opinion of the whole community.

Remarks: When the population is homogeneous i.e., when all the units are identical, it is obvious that a sample of just one element can be sufficient to provide correct information about the population.

Example: A few drops of blood can give the idea of the nature of blood in the whole body.

Main difference between Census and Sampling

Aspect	Census	Sample
Cost	More	Less
Effort	More	Less
Time consumed	More	Less
Errors	May be predicted with certain confidence level	No such errors
Accuracy of measurements	More	Less

**Source:* Advanced Sampling Theory with Applications by Sarjinder Singh

Sampling: Sampling consists of selecting a small representative part of the population.

The sampling can be performed in two ways:

1. Sampling with replacement (SWR)
2. Sampling without replacement (SWOR)

Sampling with replacement: Under this method of sampling, a unit once drawn from the population is replaced before the next draw each time.

Sampling without replacement: Under this method of sampling, the selected population units are not replaced and thus they are not eligible to be selected once again in the next draw.

Note: Practically, without a replacement sample is better than with a replacement sample.

Types of Sampling: The sampling techniques usually depend upon the nature of data and the type of enquiry. Based on these characteristics, the sampling is broadly classified into the following ways:

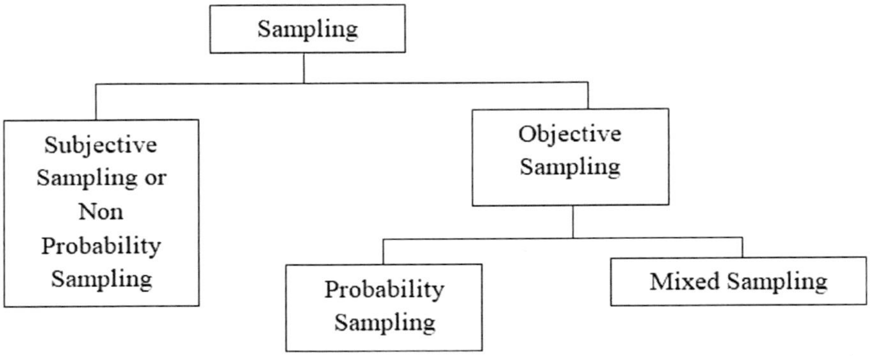

Subjective Sampling or Non-Probability Sampling

In this sampling, the selection of the units depends on the personal judgment of the investigator. The investigator may include some faulty units in the sample for his/her definite purpose in view of the desired results. The non-probability sampling produces biased results which systematically differ from the truth about the population.

Example: If an investigator wants to give the scenario that the living standard has been increased in a city say, Solan. He may take the sample of the individuals of higher income group from posh localities in the city and ignore the low-income groups and middle-class families etc.

Drawbacks: This sampling technique is often used and cannot be recommended for general purposes due to biasness of the results.

Note: If the investigator is experienced and skilled and if they carefully applied this sampling, then the non-probability sampling may yield valuable results.

Classification of Subjective or Non-Probability Sampling

1. Purposive Sampling
2. Judgment Sampling
3. Quota Sampling
4. Deliberate Sampling
5. Convenience Sampling

Merits of Non-Probability Sampling

1. This method of sampling is very easy and simple.
2. After determining the sample size with the help of planning, a suitable sample may possibly be obtained.
3. Important units or individuals may be included in the sample.

Demerits of Non-Probability Sampling

1. Predetermined view of selector affects the selection of sample which impure the results.
2. There is no place for probability in the selection of units therefore sampling error cannot be detected.
3. There is no guarantee of the validity of results from the sample selected by this method.
4. The attitude and biasness of the investigator also affect the selection of the sample that's why the results obtained by this method are not reliable scientifically.

Objective Sampling: When the sample is selected from the population with some definite rule, then sampling is known as objective sampling. The objective sampling can be further classified as:

Probability Sampling

In probability sampling, each unit in the population has some definite pre-assigned probability of being selected in the sample. There may be different types of probability sampling:

a. Every element has an equal opportunity for selection.
b. Sample units have different options for selection.
c. The probability of selection of a unit is proportional to the sampling size.

Mixed Sampling

If the sampling is partly probabilistic and partly non-probabilistic, then it is called mixed sampling.

Sampling Frame: The sampling frame is the complete list of population units from which the sample is drawn.

Examples

- For Household Survey, the sampling frame is the list of all street addresses.
- For Agricultural Survey, the sampling frame is the list of all farmers.

Under Coverage: Members/Units of the target population are excluded.

Over Coverage: Members/Units of the target population are included more than once, generally twice.

Sampling Unit: A single element or group of elements subjected to be sampled from the population is called a sampling unit. Sampling is the building block of a data set, an individual member of the population, clusters members, or some other predefined unit.

Example

- For Household Survey, the sampling unit is the individual that lives in a particular street.
- For Agricultural Survey, the sampling unit is the individual farmer.

Sampling Survey: A sampling survey is a survey, carried out on a properly chosen representative sample.

Parameter of Interest: The statistical constants of the population are called parameters. e.g., population mean, population proportions etc.

Thus, we can say that parameter is a numeric value, calculated from all the units in the population.

In a sample survey, we are mainly interested in the following parameters.

Population Total = $X = \sum_{i=1}^{N} X_i = X_1 + X_2 + \ldots + X_N$

Population Mean = $\bar{X} = \frac{1}{N}\sum_{i=1}^{N} X_i$

Population Variance = $S_X^2 = \frac{1}{N-1}\sum_{i=1}^{N}\left(X_i - \bar{X}\right)^2$

Statistic: A statistic is a function of sample observations.

Example

1. Sample Mean = $\bar{x} = \frac{1}{n}\sum_{i=1}^{n} x_i$, is a statistic based on sample data.
2. Sample proportion p = $\frac{x}{n}$ is a statistic based on x. It is the observed number of units having a specific characteristic in the sample.

Difference between Parameter and Statistic

A parameter is a fixed numerical value for an entire population. Whereas statistic is the value varies from sample to sample. Thus, the parameter is a constant value based on the entire population and statistic varies from sample to sample so that it is a random variable.

Sampling Fluctuation: The difference in the values of a statistic observed for different samples is called sampling fluctuation. In other words, since we know that statistic values vary from sample to sample. The differences between these variations are called sampling fluctuation.

Remarks: A statistic $t = t(x_1, x_2, \dots, x_n)$ is a function of random samples x_1, $x_2, \dots, x_n$ for statistical validity, the statistic t should be an unbiased estimate of population parameter θ, i.e., $E(t) = \theta$.

Sampling Distribution of Statistic: A statistic is a random variable. The probability distribution of a statistic is known as its sampling distribution.

Standard Error: The standard deviation of the sampling distribution of statistic is known as its standard error.

If t is any statistic, then for large samples:

$$Z = \frac{t - E(t)}{\sqrt{V(t)}} \sim N(0,1)$$

$$Z = \frac{t - E(t)}{SE(t)} \sim N(0,1)$$

The deviation $t - E(t)$ could have arisen due to fluctuation of sampling thus standard error measure the variability of statistic.

The reciprocal of standard error is taken as the measure of reliability or precision of the sample.

i.e., Precision of Sample $= \frac{1}{SE(t)}$

Standard Error of Statistic:

Sr. No.	Statistic	Notation	Standard Error
1.	Sample Mean	$\bar{x}$	$\frac{\sigma}{\sqrt{n}}$
2.	Sample Proportion	p	$\sqrt{\frac{pq}{n}}$
3.	Sample Standard Deviation	s	$\frac{\sigma}{\sqrt{2n}}$

Sr. No.	Statistic	Notation	Standard Error
4.	Difference between means	$\overline{x}_1 - \overline{x}_2$	$\sqrt{\frac{s_1^2}{n_1} + \frac{s_2^2}{n_2}}$
5.	Difference between proportions	$p_1 - p_2$	$\sqrt{\frac{p_1q_1}{n_1} + \frac{p_2q_2}{n_2}}$
6.	Sample Correlation Coefficient	ρ	$\frac{(1 \quad)}{\sqrt{\ }}$

Multiple Choice Questions (MCQ's)

1. Sample is considered as a part of:

   a) Population b) Data

   c) Community d) Variable

2. Sampling is inescapable in the circumstance(s):

   a) Immeasurable population

   b) Testing the success rate of COVID-19

   c) Side-effects of a drug on a community

   d) All of these

3. Which of the following is an example of primary data?

   a) Report b) Book

   c) Magazine d) Census

4. Rosy is interested in knowing the behaviour of kids in the Montessori's of Solan district; she should prefer............... method of data collection.

   a) Experimentation b) Observation

   c) Schedule d) Interview

5. Facts and figures serving as a basis for research and analysis are called:

   a) Information b) Reports

   c) Data d) Survey

6. If a researcher wants to know the impact of spray chemicals on disease control in a particular plant, then he should go for:

   a) Observation b) Survey

   c) Experimentation d) Reports

7. Recording data for each and every unit of the population is popularly known as:

   a) Sampling b) Complete enumeration

   c) An audit d) A poll

8. Which of these is the probability sampling method?

   a) Purposive b) Convenience

   c) Judgment d) Simple random sampling

9. Estimate is the term used in sampling theory for:

   a) Sampling error b) Population constant

   c) Sampling function d) Sampling method

10. Stars in the sky have............... population.

    a) Finite b) Restricted

    c) Hypothetical d) Infinite

11. An estimator is a function for estimating:

    a) A sample b) A population

    c) A parameter d) A sampling error

12. A parameter has..............value.

    a) Fixed b) Variable

    c) Constant d) Sample

13. A complete list of voters in a polling station is known as:

    a) Sample estimate b) Sampling frame

    c) Sampling distribution d) Sampling unit

14. A function of sample observations that varies from sample to sample is known as:

    a) Estimate b) Statistic

    c) Parameter d) Population

15. Larger the standard error of the sample.............. will be the precision of the sample.

    a) Smaller b) Constant

    c) Larger d) Unity

16. Process of picking up the sample from a population is known as:

    a) Survey research b) Census

    c) Auditing d) Sampling

17. The basic objective of sampling is to draw............ about the population.

a) Inferences
b) Conclusions
c) Estimates
d) Results

18. The larger the size of population, the............ should be the sample size.

a) Smaller
b) Constant
c) Larger
d) Unity

19. The measure describing any population characteristic is known as:

a) Inference
b) Estimate
c) Statistic
d) Parameter

20. The standard deviation of any sampling distribution is called as:

a) Standard error
b) Non-sampling error
c) Type-I error
d) Type-II error

21. The deviation of a statistic from a parameter is called as:

a) Probability
b) Sampling error
c) Randomness
d) Standard error

22. In probability sampling, the probability of selecting an item from the population is:

a) Known
b) Unknown
c) Undecided
d) Zero

23. Which of the following statement is true?

a) Standard deviation is always one
b) Standard deviation is always zero
c) Standard deviation is always positive
d) Standard deviation is always negative

24. A complete scheme for obtaining a sample from a population is called:

a) Population design
b) Sampling design
c) Sampling frame
d) Sampling plan

25. Sampling distribution has........... shape.

a) Binomial
b) Flat
c) Positively skewed
d) Normal

26. Which of the following sampling method is not based on probability?

a) Cluster
b) Simple random sampling
c) Quota
d) Systematic

27. Choose the correct statement in regards to sampling with replacement:

a) Repeated selection of single unit
b) Formation of infinite population
c) Sample size may be greater than population size
d) All of these

28. Value of OGPA of the students enrolled in a university in 2021is:

a) Infinite
b) Finite
c) Unlimited
d) Boundless

29. The errors other than sampling errors are termed as:

a) Type-I errors
b) Non-sampling errors
c) Type-II errors
d) Standard error

30. Which one of the following is not a potential benefit of sampling?

a) Cost saving
b) Time-saving
c) More accurate than a complete enumeration
d) Involves fewer efforts

31. Choose the incorrect statement about non-probability sampling:

a) Subjective in nature
b) More accurate
c) Easy and simple
d) Purposive in nature

32. The sampling technique involving both probabilistic as well as non-probabilistic components is known as:

a) Subjective sampling
b) Mixed sampling
c) Systematic sampling
d) Objective sampling

33. Individual farming household in a socio-economic survey is known as:

a) Sampling unit b) Variable

c) Estimate d) Sampling frame

34. The alternative term for random sampling is:

a) Probability sampling b) Non-probability sampling

c) Sampling error d) Random error

35. The other term for non-random sampling is:

a) Biased sampling b) Non-probability sampling

c) Random sampling d) Representative sampling

Key to MCQ's

1.	(a)	2.	(d)	3.	(d)	4.	(b)	5.	(c)
6.	(c)	7.	(b)	8.	(d)	9.	(b)	10.	(d)
11.	(c)	12.	(b)	13.	(b)	14.	(b)	15.	(a)
16.	(d)	17.	(a)	18.	(c)	19.	(d)	20.	(a)
21.	(b)	22.	(a)	23.	(c)	24.	(b)	25.	(d)
26.	(c)	27.	(d)	28.	(b)	29.	(b)	30.	(c)
31.	(b)	32.	(b)	33.	(a)	34.	(a)	35.	(b)

Let us do

1. In agricultural sciences, is there any difference in the level of their reliability and if so, why?
2. In the statements given below, specify the type of data either primary or secondary?
   a) Size of land holdings of the farmers of Himachal Pradesh published in state statistical abstracts.
   b) Data were collected by a research scholar on the cropping pattern of crops on their land.
   c) Financial statement of Adani group of industries.
   d) Data collected by a researcher on timber export from India from FAO STAT website.
3. Define sample. Why we do sampling?
4. Distinguish between sample and population by giving suitable examples.

5. Differentiate between parameter and estimate.
6. Differentiate between finite and infinite population.
7. Distinguish between real and hypothetical population.
8. Differentiate between parameter and statistic.
9. Discuss various methods of primary data collection in brief by citing suitable examples. Also, discuss their advantages and disadvantages.
10. Give some circumstances under which taking up sampling is irresistible.
11. Suggest the suitable technique either sample survey or complete enumeration in the following circumstances:
    a) A researcher has to work out the agricultural mechanization index of the district or state.
    b) Population demographic features.
    c) Economic analysis of dairy farming in Himachal Pradesh.
12. Differentiate between standard error and standard deviation.
13. Differentiate between sampling and non-sampling errors.
14. The state department of horticulture wants to know the willingness of the apple growers to pay for weather-based crop insurance schemes in Himachal Pradesh. Discuss in brief the procedural steps to plan a sample survey for the cause.

2

Sample Design

Sample Design

Sample design is a specific system for obtaining a sample from a given population. For the effective planning and sample survey, this sample design is the framework or road map that serves as the basis for the selection of the survey sample and affects many other important aspects of a survey as well. It is a definite plan for obtaining a sample from a given population. Sampling design refers to the procedure that the researcher would adopt in selecting items as the sample. Also, sampling design is a working plan that specifies the sample frame, sample size, sample selection and estimation method in detail. The blueprint for obtaining a sample from the sampling frame and this technique used in drawing a sample is known as sample design or sampling design. This whole process is done before the determination of data collection.

Principle Steps in Sample Survey

The following steps are important to make the process most effective.

1. **Objective of Survey:** The first step of sampling design is to define the objectives of the survey in a clear and concrete manner. The sponsors or investigators of the survey should take care that the proposed objectives and the resources available to fulfil the objective *viz.* money, manpower and time limit etc. must be commensurate with one another.

2. **Sampling population:** In order to meet the objectives of the survey, the population should be clearly defined. To avoid any uncertainty regarding the coverage of the survey, restrictions in terms of geological, physical and other parameters may be avoided.

Defining Sampling Unit and Sampling Frame

Sampling Unit: Usually the sampling unit is decided in such a way that it must represent the whole population. While defining the sampling units it becomes important to define individual units as a distinct, unambiguous and non-overlapping population group. It may be categorized on the basis of a geographical unit or a construction unit or a social unit or it may be an individual unit.

Geographical Unit: State, District, Block, Village etc.

Construction Unit: House, Flat etc.

Individual Unit: Ram, Shyam, Reeta, Geeta or any individual person.

Sampling Frame: In order to cover the entire population through a representative sampling unit, there should be a list, map or other acceptable material called a sampling frame. It is extremely important for the sampling unit, to be in the sampling frame.

3. **Data to be collected:** The objective of the sample survey should be much clear to the investigator. He should keep in mind that no irrelevant information should be collected and no essential information should be discarded.
4. **Pilot Survey:** Before the field execution of the sampling design, a thorough field survey may be necessarily performed on a small scale to increase the efficacy of the model. This is called a pretest or pilot survey and has a special consideration in the modelling for giving better insight into the practical problems which may arise later during the execution of the actual model in the field.
5. **Field Work:** A successful survey depends on reliable fieldwork for collecting data, there should be efficient supervisory staff and trained persons for conducting fieldwork.
6. **Summary and Analysis of Data:** This usually is the last step in the model design wherein inferences are drawn based on the information provided in the collected data. For an effective summary and data analysis following steps must be taken care of:
   i. Scrutiny of the filled questionnaires to work out the probable and consistent nature of the data.
   ii. A hand or machine tabulation of the data is a necessary component.
   iii. Once the data scrutiny, editing and tabulation are done. A careful, precise and need-based statistical analysis of the data is to be done keeping in view the findings desired.
   iv. Once the suitable statistical analysis has been done the report representing the detailed statement of the procedure followed and the findings obtained should be prepared for reference to the concerned.

This entire process completes the basic procedure followed for the effective sampling design.

Basic Principles of Sample Survey

1. **Statistical Regularity:** The term statistical regularity implies to a moderately large number of samples randomly selected from a population that on an average should possess all the characteristics of the large group of population. This means that the samples must be selected in a way that they shall represent the entire population as a whole. This justifies the need and significance of the sample selections at random intervals and instances to provide every unit of the subject/population an equal opportunity of being selected as a representative unit of the population.
2. **Validity:** The sample should be so selected that valid estimators are available and their properties (like unbiasedness, minimum variance etc.) can be studied objectively. This principle should be satisfied by a probability sample.
3. **Optimization:** The process of optimization takes into consideration for the efficiency of the sampling procedure and the ultimate cost of a survey. Further, the efficiency of an estimator is measured by the inverse of its sampling variance, while the cost is generally measured in terms of expenditure incurred on the survey.

From this inference can be drawn that the optimization ensures:

- A fixed level of efficiency can be achieved for a minimum cost.
- A maximum level of efficiency can be achieved for a fixed cost.

Errors in sample survey

Error is defined as the tendency to exclude or include certain units or individuals in the whole process of the sample survey. The error is involved in various steps of the sampling including data collection, its processing/tabulation and ultimate in the statistical analysis. It is broadly classified into two categories.

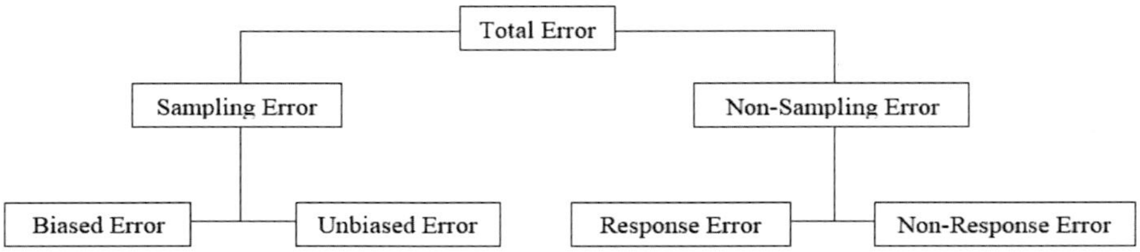

Sampling Error: The error, that originated only from sampling is called sampling error. It occurs when the investigator does not select a sample that represents the entire population of data. There are various factors which can affect the sampling error while the most important ones include the sample size and design, sampling fraction, variability within the population etc.

Sampling errors arise from the fact that only a part of the population is used for estimation. So, whatever may be our degree of cautiousness in selecting a sample and in recording its details, there will always be a difference between a parameter and its corresponding estimate obtained from sample values. The different samples taken from the same population under identical conditions are also expected to give different estimates. The difference between the parametric value and its estimates derived from the sample is called sampling error.

Key Notes

- Sampling errors are absent in complete enumeration.
- Sampling errors are inherent and unavoidable in any sampling scheme.
- With the increase in sample size and by adopting a suitable sampling scheme, we can reduce our sampling error.

Biased Sampling Error

When the selection of the sample is based on the personal prejudice or bias of the investigator then the results are prone to bias errors. Such as, if the investigator is required to collect the sample by using simple random sampling and instead, he used non-random sampling, then personal prejudice is introduced in the research process that will lead to biased errors.

Unbiased Sampling Error

These errors arise due to chance, i.e., the investigator has not intentionally tampered with the sample and the difference between the population and sample has occurred by chance. The differences between the members of the population included in the sample and those who are not included in the sample.

Non-Sampling Error

There may be different sources for the non-sampling error to occur while it can arise at any stage of observation, ascertainment and data processing. This is further certain that the non-sampling error may be present in both the census as well as the sample survey and can arise at every phase of the planning or execution.

The following factors may result in the non-sampling error:

- The objectives of the survey need to be stated clearly, if not this results in the incorporation of the samples not desired and the exclusion of the samples which are necessary to be incorporated in the sample.
- If the data specifications provided are inadequate and inconsistent in relation to the proposed objectives.
- The ill-designed questionnaire may also result in errors.
- Absence of trained and well-qualified investigators and supervision.
- Errors may arise due to incorrect responses from the respondents.
- Errors may also arise during the processing of the data due to faulty editing, coding, tabulation and summarization of data.
- The non-sampling error may also occur during the presentation and printing of the result.

Comparison	Sampling error	Non-sampling error
Meaning	This error occurs due to the sample selected does not perfectly represent the population of interest.	This error occurs due to sources other than sampling while conducting survey activities.
Cause	Deviation between sample mean and population mean.	Deficiency and analysis of data.
Occurs	Only when sample is selected.	Both in sample and census.
Type	Random	Random and Non-random
Sample size	With the increase in sample size there is a possibility of decrease in error.	It has nothing to do with the sample size.

Advantages of Sampling over Census

Following are the main reasons for preferring sampling over the census.

1. **Reduction of Cost:** Sampling often lowers the costs in terms of financial and human resources. The total cost for the sampling is however expected

to be lower compared to the census. If a population is very large, the census can be very expansive.

2. **Greater speed:** Using a sample, the data can be collected more quickly so that the estimates can be published in time. In the census, the collection of data may take so long time.
3. **Greater Scope:** A complete enumeration or census requires a large administrative setup and involves many persons in data collection. Therefore, a sample survey has a greater scope than a census survey. Some enquiries may require highly trained personnel or specialized equipment for the collection of data, which sometimes may make census unmanageable.
4. **Greater Accuracy:** Estimates, based on a sample are often more accurate than those based on the census because investigators can be more careful while collecting data for fewer units. With the huge administrative complexity and the pressure to produce estimates timely, many types of errors can easily creep into the census data.
5. **Greater Quality:** A sample survey assists in a better quality of the data compared to the complete census as in the sample survey, there is a possibility to use better resources precisely.
6. **Measure of Precision of Estimates:** In complete enumeration, there is no way of gauging the magnitude of errors incorporated in the estimates. But a sample drawn based on a properly designed survey method permits quantitative assessment of the errors involved in the estimates.
7. **Greater Applicability:** In some studies, observations are obtained by destroying the units.

For example: To obtain the average life hour of electric bulbs, we have to measure the total hours that the bulbs survive until they are burnt out. In such cases, drawing of sample is essential since complete enumeration of the bulbs destroys the entire population.

*However, when it is essential to collect information from all the units of population irrespective of cost or time, a complete census is required. For example, **"population census"** where all relevant information is collected from all the units of population.*

Multiple Choice Questions (MCQ's)

1. The roadmap for selection of the survey sample is known as:

   a) Sample plan b) Sample design

   c) Sample distribution d) Sample frame

2. A sample frame involves the specification of............ characteristics during sampling.

   a) Error frame b) Random frame

   c) Constant frame d) Population frame

3. A thorough field survey on a small scale performed usually before the execution of sampling design is:

   a) Pilot survey b) Evaluation

   c) Field survey d) Sample survey

4. .... is the first step in sample designing.

   a) Defining the population b) Sampling objectives

   c) Pilot survey d) Summary and data analysis

5. Sampling unit must possess the property of:

   a) Symmetrical b) Asymmetrical

   c) Overlapping d) Non-overlapping

6. The final step in sample designing involves drawing inferences from the available data is:

   a) Fieldwork b) Summary and data-analysis

   c) Sample population d) Sample selection

7. The inferences drawn from a study of young single mothers at the college can be summarized to the population of:

   a) All young single mothers at the college

   b) All single mothers at the college

   c) All young single mothers in the society

   d) All young single mothers in the community

8. Which of these is not a principle of sampling survey?

   a) Principle of validity
   b) Principle of statistical regularity
   c) Principle of optimization
   d) Principle of standardization

9. Sampling error is absent in which of the following circumstance(s).

   a) Sampling
   b) Complete enumeration
   c) Household survey
   d) Village survey

10. The principle of sample survey design that focuses on a truly representative sample of the target population is:

    a) Principle of statistical regularity
    b) Principle of optimization
    c) Principle of validity
    d) Principle of standardization

11. Larger the sampling variance of the estimator, the............ will be the efficiency.

    a) Constant
    b) Smaller
    c) Larger
    d) Unity

12. Sampling errors cannot be minimized by:

    a) Increase in sample size
    b) Proper sample design
    c) Selecting a representative sample
    d) Decrease in sample size

13. A complete list of blood donors in a blood donation camp is known as:

    a) Sample estimate
    b) Sampling frame
    c) Sampling distribution
    d) Sampling fluctuation

14. If an electronic company desired to assess the life of tube lights manufactured of a new design, the company should prefer:

    a) Sample study
    b) Complete enumeration
    c) Company's current income
    d) Company's previous income

15. The target population cannot be defined in terms of:

    a) Time period under consideration
    b) Sampling units available for selection
    c) Geographical area under consideration
    d) Sampling units which are not available for selection

16. An institute wishes to determine the adoption rate of a subscribed insurance scheme amongst a faculty of 200.

    a) Survey research b) Census
    c) Auditing d) Sampling

17. The type of sampling errors that arises due to personal bias in the sampling process is:

    a) Biased errors b) Unbiased errors
    c) Type-I errors d) Type-II errors

18. The reason(s) for the occurrence of non-sampling errors in our research is:

    a) Faulty planning b) Faulty selection of sample
    c) Lack of managerial skills d) All of these

19. Non-sampling errors are more common in:

    a) Census b) Sampling survey
    c) Both (a) & (b) d) None of these

20. The main advantage of sampling over census is:

    a) Greater speed b) Reduced cost
    c) Greater scope d) All of these

Key to MCQs

1.	(b)	2.	(d)	3.	(a)	4.	(b)	5.	(d)
6.	(b)	7.	(a)	8.	(d)	9.	(b)	10.	(a)
11.	(b)	12.	(d)	13.	(b)	14.	(a)	15.	(d)
16.	(b)	17	(a)	18.	(d)	19.	(c)	20.	(d)

Let us do

1. Define sampling design. Discuss its process in brief.
2. Discuss various principles of statistical survey design? What are the possible after-effects of not following these principles during the sample survey design?
3. Define the pilot survey. What are the potential difficulties a research scholar may experience if he skipped the pilot survey?
4. Prepare a sample survey design under the research problems:

a. Impact of MGNREGA on its beneficiaries in India.

b. Economic analysis of Mushroom cultivation in mushroom city Solan.

c. Impact of growing Pansy flower on the mental health of students and staff in a university.

d. Psychological well-being of bank employees in Delhi.

5. Discuss in brief the significance of fixing an appropriate sample size while designing a sample survey?

6. Write in short about the following concepts along with suitable examples:

   a) Unbiasedness

   b) Random sample

   c) Optimization

   d) Confidence interval

   e) Response Vs non-response errors

   f) Sampling unit

7. What do understand by the term "Non-sampling errors"? What are their main drivers in a research journey?

8. Give the type of sampling error that may occur in a sample survey under the following incidences (Sampling/Non-sampling):

   a) If a researcher chooses his hometown and own village for a survey.

   b) Wrongly specified population.

   c) Misleading statistical analysis and inferences drawn.

   d) Improper sampling technique applied.

   e) Inadequate sample size.

3

Simple Random Sampling

Simple Random Sampling (SRS)

Simple random sampling is probability sampling and refers to a technique of sampling where each and every unit of the population have an equal and independent opportunity of being selected and included in the sample during the sampling. In the SRS the selection of the sample depends on the chance factor and is also known as the method of probability sampling.

There are two ways in which SRS can be performed:

- **SRSWR:** Simple Random Sampling With Replacement
- **SRSWOR:** Simple Random Sampling Without Replacement

SRSWR: In the SRSWR, if the samples are selected one by one and returned to the population before the next drawing, the sampling procedure is known as SRSWR.

Let sampling with replacement is done to draw a possible sample number of a size n from the random population of N size.

The first unit of the sample may be drawn from N units in ${}^{N}C_{1} = N$ ways.

Similarly, there are N ways to draw the second unit from the population because as the 1st selected unit is returned to the population and so on up to the selection of nth unit.

But this can be hypothesised that the total number of ways to draw the sample from the population is ${}^{N}C_{1} \cdot {}^{N}C_{1} \cdot {}^{N}C_{1} \ldots {}^{N}C_{1}$ (n times) that is $({}^{N}C_{1})^{n} = N^{n}$ ways.

Therefore, the probability of each possible sample being selected with SRSWR is:

$$p = \frac{1}{N^{n}}$$

Example

Let us consider a population of size 3 i.e., N = 3, as A, B, C and we want to draw a sample of size 2 i.e., n = 2.

If we choose SRSWR to draw the sample of size 2 then, all the possible samples are N^n i.e., $3^2 = 9$ samples and they are $S_1 = (A, A)$, $S_2 = (A, B)$, $S_3 = (A,C)$, $S_4 = (B, A)$, $S_5 = (B, B)$, $S_6 = (B, C)$, $S_7 = (C, A)$, $S_8 = (C, B)$, $S_9 = (C, C)$ out of these 9 samples there is a possibility to get select a sample of size 2 is either $S_1 = (A, A)$ or maybe any one of the rest samples.

Here, in that case, every unit has the same probability to get selected out of A, B and C i.e., $\frac{1}{3}$ and every sample of size n = 2 has an equal probability to get selected i.e., $\frac{1}{9}$

SRSWOR: Whenever the simple random sample is selected one by one in such a way that the units are not returned to the population before the next drawing, the sampling procedure is known as SRSWOR.

Suppose an arbitrary sample size of n from the given population of the size N is selected with SRSWOR.

The probability of selecting the units $u_1, u_2, u_3, \ldots, u_n$ in the order:

$$u_1 \to u_2 \to u_3 \to \ldots \to u_n \text{ is } \frac{1}{N} \times \frac{1}{N-1} \times \frac{1}{N-2} \times \ldots \times \frac{1}{N-(n-1)}$$

Since these n units can be selected in n! ways.

Thus, the probability of each possible sample being selected with SRSWOR is:

$$p = n! \times \frac{1}{N} \times \frac{1}{N-1} \times \frac{1}{N-2} \times \ldots \times \frac{1}{N-(n-1)} \Rightarrow p = \frac{1}{NCn}$$

Example

Here, population size N = 3 as A, B, C and sample size n = 2 but the sample is to be drawn with the help of SRSWOR. Then, the possible samples are $S_1 = (A, B)$, $S_2 = (A, C)$, $S_3 = (B, C)$

The probability of each possible sample is to be selected is $= \frac{1}{NCn} = \frac{1}{3}$

Remarks

- An SRSWR sample may contain a particular population unit more than once, but a particular population unit may appear more than once in a SRSWOR sample. In that sense, with replacement sampling has an effect of generating an infinite population.

Example

In with replacement sampling, there is nothing to stop one from drawing a sample of size 10 from a population of size 100.

- The probability p of each possible sample being selected is:

$$p = \frac{1}{N^n} \text{, in case of SRSWR}$$

$$p = \frac{1}{NCn} \text{, in case of SRSWOR}$$

Selection of Random Sample / Simple Random Sample

How to minimize the relevance of biases and errors in the selection process for the random sample?

Random sampling is the method of sampling where each item has an equal chance of selection. However, random sampling is not entirely dependent on the selection method rather the size and nature of the population also affect the procedure. Generally, the method of selection should be independent of biases and errors. The random samples can be acquired by any of the following selection methods:

- Lottery method or Chit method
- Pack of cards
- Random number method or Mechanical Randomization method

Lottery method or Chit method

This method is the most primitive and mechanical example of random sampling in which:

- All population items are listed or named in separate papers.
- The slips need to be of the same size, colour and shape.
- These slips are folded, mixed up and shuffled thoroughly.
- Then the selection is then made based on the desired sample size.

The selection by this procedure thus is entirely dependent on the chance factors and the sample drawn is random and free from biases.

Example

To follow the selection of n candidates out of the given N size of the population numbers from 1 to N are assigned randomly to the population. These numbers now correspond to N number of slips which identifies each individual while attempts are made to make the group as homogeneous as possible. The slips (representing the individuals) are placed in a container and thoroughly shuffled and ultimately n slips are drawn one by one and the n candidates, who correspond to the numbers drawn through the slip may be randomly selected giving all the candidates an equal opportunity to get selected.

SRSWR and SRSWOR in Lottery Method

- As soon as a slip containing any number is drawn from the container and after noting down the number it is placed back in the container before the next draw commences, the number of slips which represents the requisite sample size is referred to as the SRSWR sample.
- As soon as a slip containing any number is drawn from the container and after noting down the number it is not put back in the container for the next draws, then it's far referred to as SRSWOR.

Pack of Cards

The pack of cards is a kind of miniature of the population for sampling purposes. We make one card correspond to one of the units of the population by writing on it the number of units. The cards are then shuffled a number of times and then a card is drawn at random from them. This is the most reliable method of selecting a random sample.

Random Number method or Mechanical Randomization method

If the population is sufficiently large, the first two methods of sampling viz. the lottery method and pack of cards method are quite time-consuming and cumbersome tasks. An alternative method of random selection is that of using the random number table, which has been so constructed that each of the digits 0 to 9 appears with approximately the same frequency.

If we have to select a sample from a population of size N ($\leq$ 99 i.e., two digits population size) then the numbers can be combined two by two to give pairs from 00 to 99. Similarly proceed for three digits, four digits and so on.

How to draw a random sample contains the following steps.

a) Identification of the N units in the referred population, and their numbering from 1 to N.

b) Random selection of the samples, any page, random tables and random numbers in any row or column or diagonal and discard any values greater than N.

c) The sample units represent the required numbers, preferred in step b comprise the random sample.

Theorems based on Simple Random Sampling

Theorem 1: In SRSWOR

i. The probability of drawing / selecting a specified unit at the rth draw is equal to the probability of drawing it at the first draw.

   i.e., $P(E_r) = P(E_1)$

ii. The probability of including a specified unit in the sample of size n is $\frac{n}{N}$

Proof

i. Let E_r be the event that any specific unit is selected at the rth draw.

   Therefore, (E_r) = P (The specified unit is not selected in the first (r-1) draw and then selected in the rth draw)

$$= \prod_{i=1}^{r-1} \left[1 - \frac{1}{N-(i-1)} \right] \times \frac{1}{N-(r-1)}$$

$$= \prod_{i=1}^{r-1} \left[1 - \frac{N-i}{N-(i-1)} \right] \times \frac{1}{N-(r-1)}$$

$$= \left[\frac{N-1}{N} \times \frac{N-2}{N-1} \times \ldots \times \frac{N-r+1}{N-r+2} \right] \times \frac{1}{N-r+1}$$

$$\Rightarrow P(E_r) = \frac{1}{N} = P(E_1)$$

   = Probability of selecting any unit at the first draw.

ii. Since a specified unit can be included in the sample of size n at any of n draws. The required probability is the sum of the probabilities of n mutually exclusive events namely the unit is included in the sample at the first draw, second draw, . . . , n^{th} draw.

   The required probability $= P(E_1) + P(E_1) + \ldots + P(E_n)$

$$= \frac{1}{N} + \frac{1}{N} + \ldots + \frac{1}{N} = \frac{n}{N}$$

   Probability $= \frac{n}{N}$ Hence Proved

Theorem 2:

In SRSWR, the selection of up to n units (one by one), can be done in N^n ways.

Where, n = sample size and N = population size

The first unit can be drawn from N units in ${}^NC_1$ ways and the second unit can also be drawn in ${}^NC_1$ ways because the first selected unit is again mixed with the population and so on up to the nth unit.

Thus, the total number of ways:

$= {}^NC_1 \,.\, {}^NC_1 \,.\, {}^NC_1 \,.\, \ldots \,.\, {}^NC_1$ (up to n^{th} unit)

$= ({}^NC_1)^n = N^n$

Theorem 3:

The sample mean $\bar{y}$ is an unbiased estimator of population mean $\bar{Y}$ in both the cases (i) SRSWR and (ii) SRSWOR.

Proof

Case (i) SRSWR

We have, $\bar{y} = \frac{1}{n}\sum_{i=1}^{n} y_i$ .... (1)

Taking expectations on both sides, we have

$$E(\bar{y}) = E\left(\frac{1}{n}\sum_{i=1}^{n} y_i\right) = \frac{1}{n}\sum_{1}^{n} E(y_i) = \frac{1}{n}\sum_{i}^{n} \bar{Y} = \frac{1}{n} \times n\bar{Y} = \bar{Y}$$

$\Rightarrow E(\bar{y}) = \bar{Y}$

Thus in SRSWR sample mean $\bar{y}$ is an unbiased estimator of population mean $\bar{y}$

Case (ii) SRSWOR

We have $\bar{y} = \frac{1}{n}\sum_{i=1}^{n} y_i = \frac{1}{n}\sum_{i=1}^{N} a_i Y_i$ .... (2)

Where, $a_i = \begin{cases} 1 \\ 0 \end{cases}$ if i^{th} unit is included in the sample, 0 otherwise

Taking expectations on both sides of (2), we get

$$E(\bar{y}) = \frac{1}{n}\sum_{1=1}^{N} Y_i E(a_i) \quad \ldots\ldots\ldots\ldots \quad (3)$$

Since a_i takes only two values 1 and 0,

$\therefore E(a_i) = 1 \cdot P(a_i = 1) + 0 \cdot P(a_i = 0)$

$= 1 \cdot P$ (i^{th} unit is included in the sample) $+ 0 \cdot P$ (i^{th} unit is not included in the sample) $= 1 \cdot \frac{n}{N} + 0 \cdot \frac{n}{N} = \frac{n}{n}$ From Theorem 1(ii)

So, from equation (3), we have

$$E(\bar{y}) = \frac{1}{n} \times \frac{n}{N} \sum_{i=1}^{N} Y_i = \frac{1}{N} \sum_{i=1}^{N} Y_i = \bar{Y}$$

$$\Rightarrow E(\bar{y}) = \bar{Y}$$

Thus, in SRSWOR, the sample mean $\bar{y}$ is an unbiased estimator of population mean $\bar{y}$

Determination of Sample size

In sampling theory, some of the most important problems a statistician or researcher may face before planning for the sample survey are:

i. The sample size may be defined in detail.

ii. Size of the sample (large or small) that may be representative of the whole population.

iii. Whether the estimated sample consists of the smallest sampling error.

iv. How to determine the sample size for the statistical study.

Two Important Facts

1. If we take too small sample size, it may not be representative of the entire population and may not serve to achieve the objective of the study.
2. If sample size is too large, it may acquire huge money, time and human resources.
   - The sample which has the lowest sampling error is considered a good representative of the sampling population.
   - Sample must be computed by using some statistical procedures to minimize these errors.

Basic factors for determining sample size

There are three statistical considerations in determining the appropriate sample size:

1. Degree/level of precision
2. Desired confidence level
3. Degree of variability

Degree or level of precision

- Level of precision is defined as the "Margin of Error" between the estimate of sample value obtained and the given population value.
- It is also known as "Margin of Permissible Error" or "Sampling Error".
- It is based on the proportion of risk a researcher or statistician is willing to tolerate in dealing with the available data to make precise decisions.
- The margin of error is generally expressed as a percentage.

Margin of Error (%)	2 %	3%	4%	5%	10%
Sample Size	2401	1067	600	384	96

High level of precision (Lower level of Margin of Error) requires more wide spectrum sample selection and thus may induct higher cost to achieve the desired sampling size.

- While computing the samples, the intended confidence level is defined by Z value.
- By using Central Limit Theorem, we know

  If $X \sim N(\mu, \sigma^2)$ then $\bar{x} \sim N\left(\mu, \frac{\alpha^2}{n}\right)$ and thus $Z = \frac{x - \mu}{\sigma} \sim N(0,1)$

Desired confidence level

Area property of the normal curve:

i. $P(\mu - \sigma \leq x \leq \mu + \sigma) = 68.27\%$

ii. $P(\mu - 2\sigma \leq x \leq \mu + 2\sigma) = 95.45\%$

iii. $P(\mu - 3\sigma \leq x \leq \mu + 3\sigma) = 99.73\%$

- The most usual confidence levels defined in various sectors of the scientific evaluation include 99%, 95% and 99% but any confidence level may be established.
- 99% of the area of the normal curve falls within the limits $\pm 2.58\,\sigma$ (here 2.58 is the Z score)
- 95% of the area of the normal curve falls within the limits $\pm 1.96\,\sigma$ (here 1.96 is the Z score)

Degree of variability

The 'degree of variability' describes the distribution of the attributes in the population.

- If the population group is more homogeneous, a smaller sample size may be required to draw.
- If the given population group is heterogeneous, larger sample size may be required to be drawn to attain the required level of precision (or margin of error).

An α-level confidence interval for the population proportion P is given by:

$$\hat{p} - e < P < \hat{p} + e \text{ where } e = Z_\alpha \sigma_{\hat{p}} = Z_\alpha \bullet \sqrt{\frac{\hat{p}(1-\hat{p})}{n}}$$

$$\text{i.e., } e = Z_\alpha \bullet \sqrt{\frac{\hat{p}(1-\hat{p})}{n}} = Z_\alpha \bullet \sqrt{\frac{\hat{p}\hat{q}}{n}} \text{ where } \hat{q}(1-\hat{p})$$

The value of e is known as the Margin of Error or the Error of Estimate.

Methods to determine representative sample size

- SRS is the most common and simplest method adopted for sampling.
- It draws n units of sample out of a population of size N and thus provides an equal opportunity/probability of selection to all the units of the population.
- We will discuss here the following two methods to determine sample size based on simple random sampling:
    i. Cochran's Formula (for finite and infinite population)
    ii. Yamane's Formula

Cochran's formula

The Cochran formula for calculating the ideal sample size is based on:

a) Desired level of precision

b) Desired confidence level

c) Estimated proportion of the attribute present in the population.

Cochran (1977) provided the formula for the both infinite and finite populations.

Case-1 When the population is infinite

$$n = \hat{p}\hat{q} \times \left(\frac{Z_\alpha}{e}\right)^2$$

Where

$\hat{p}$ = Population, defined from a preliminary sample $\Rightarrow \hat{q} = (1-\hat{p})$

Z_α = Z-score for the desired confidence level

e = Desired margin of error

Example

After the survey of the residents of around 240 countries based on their opinion on the changes in the election policies of their countries, around 75.80% were reported to be in the favor of the change. After using a 95% confidence level, the margin of error was reported to be more than 5% for the parameter under the study. Based on this assumption we need to reduce the margin of error to about 3%. To achieve this how many more residents are to be identified and included in the survey for the reduction of the margin of error to 3%.

Solution

At first, 240 samples have already been drawn from the population in the preliminary survey.

Margin of error e = 5% = 0.05

$\hat{p} = 75.80\% = 0.758 \Rightarrow \hat{q} = 1 - \hat{p} = 0.242$

$Z_\alpha = 1.96$

Thus, the required sample size with a new margin of error e* = 3% = 0.03

$$n = \hat{p}\hat{q} \times \left(\frac{Z_\alpha}{e^*}\right)^2 = 0.758 \times 0.242 \times \left(\frac{1.96}{0.03}\right)^2 = 783$$

We see that, for a 95% confidence level and 3% margin of error, the required sample size is 783. Thus, 783−240 = 543 additional samples should be included to reduce the margin of error up to 3%.

<u>**Important Note**</u>

- $\hat{p}$ and $\hat{q}$ are the two positive fractions such that $\hat{p} + \hat{q} = 1$
- Consider that we have a large population with an unknown degree of variability. In this case, we may assume the maximum level of variability in the population proportion, which is equal to 50%.

- Thus, $\hat{q} = 1 - \hat{p} = 0.5$, with a maximum value of 0.25 for their product. Thus, the required sample size is:

$$n = 0.25 \times \left(\frac{Z_\alpha}{e}\right)^2 = \left(\frac{Z_\alpha}{2e}\right)^2$$

Example

If you need to poll voters to determine whether they favour a ballot proposition. You need a 95% confidence level and a margin of error of 3% or less. How many voters should you include in the study?

Solution

$Z_\alpha = 1.96$ and $e = 3\% = 0.03$

Here, the measure of variability is not known, so we can assume the maximum variability which is equal to 50% i.e., $\hat{p} = 0.5 \Rightarrow \hat{q} = 1 - \hat{p} = 0.5$ So that $\hat{p}\hat{q} = 0.25$

The sample size is then:

$$n = 0.25 \times \left(\frac{Z_á}{e}\right)^2 = 0.25 \times \left(\frac{1.96}{0.03}\right)^2 = 1067.11$$

Sample size calculated for 95% and 99% confidence levels and different margin of error

Confidence Level	Z_α	Sample Size (n)			
		e = 3%	e = 5%	e = 7%	e = 10%
95%	1.96	1067	384	196	96
99%	2.58	1849	666	340	166

Case-2 When the population is finite

- According to Cochran (1977), if the population is finite, the sample size can be reduced slightly.
- This is due to the fact that a very large population provides proportionally more information than that of a smaller population size.

The formula for calculating the appropriate sample size is:

$$n = \frac{n}{1 + \frac{(n-1)}{N}}$$

This is called the correction formula for obtaining the final sample size from a finite population.

Here n_0 is the sample size obtained on the basis of formula for infinite population.

$$n_0 = \hat{p}\hat{q} \times \left(\frac{Z_\alpha}{e}\right)^2$$

If $\frac{n_0}{N}$ is negligible then n_0 is a satisfactory approximation to the sample size.

Concept

Suppose we want to estimate the sample size of a population $N = 13191$. In the above table we see that with a 5% margin of error, and at a 99% confidence level, the sample size is 666.

Thus, $\frac{n_0}{N} = \frac{666}{13191} = 0.05(\text{i.e.}5\%)$

The sample size exceeds with 5% of the population size 13191.

so that, $n = \dfrac{n_0}{1+\dfrac{(n_0-1)}{N}} = n = \dfrac{666}{1+\dfrac{(666-1)}{13191}} = 634.03 = 634$

Thus, for the population N = 13191, the sample size should be 634 at a 99% confidence level and 5% margin of error.

Yamane's formula

Yamane (1967) suggested a simplified formula to calculate sample size from a population.

- Assuming the maximum level of variability which is equal to 50% (i.e., $\hat{p} = 0.5$) the sample size for a 95% confidence level should be:

  $$n = \frac{N}{1 + N(e)^2}$$

Where, N is the population size and e is the tolerable margin of error.

Multiple Choice Questions (MCQ's)

1. The type of sampling where every unit in the universe has some probability/ chance of being selected in the sample is:

   a) Non-probability sampling　　b) Probability sampling

   c) Mixed sampling　　d) Sequential sampling

2. The type of probability sampling in which each unit in the population has some equal and independent chance of being selected is:

   a) Systematic sampling　　b) Cluster sampling

   c) Simple random sampling　　d) Stratified sampling

3. Simple random sampling can be formed in terms of:

   a) Simple random sampling with replacement

   b) Simple random sampling without replacement

   c) Both (a) and (b)

   d) None of these

4. The type of simple random sampling involving each selected sample at every draw to be replaced to the population before the succeeding draw:

   a) Simple random sampling without replacement

   b) Simple random sampling with replacement

   c) Mixed sampling

   d) Stratified sampling

5. If a researcher has to draw a sample of size (n) out of a population (N), several ways to draw the sample are:

   a) N　　b) N^n

   c) N^{n-1}　　d) n^N

6. Consider a situation, where a researcher performs a procedure100 times and in 90 out of 100 cases his results were true while 10 times the results were wrong, then the perceived level of confidence is:

   a) 95%　　b) 90%

   c) 99%　　d) 98%

7. The probability of selection of a sample in SRSWR is:

a) $\frac{1}{N^n}$ b) $\frac{1}{N}$

c) $\frac{1}{NCn}$ d) 1^{n-1}

8. The probability of selecting a sample in SRSWOR:

a) $\frac{1}{NCn}$ b) $\frac{1}{N}$

c) $\frac{1}{N_n}$ d) $\frac{1}{N_i}$

9. The probability of being chosen is non-constant at each succeeding draw is:

a) SRSWR b) SRSWOR

c) Stratified sampling d) Mixed sampling

10. Which of the following method is suitable for drawing a random sample?

a) Pack of card b) Mechanical randomization

c) Chit method d) All of these

11. The most primitive method of random selection is:

a) Mechanical randomization b) Roulette method

c) Chit method d) Cochran's method

12 .... is the probability of selection of successive units in a sample size (n) from N population units:

a) $\frac{1}{N}$ b) $\frac{n}{N}$

c) $\frac{1}{n}$ d) $\frac{\frac{1}{N}}{n}$

13 .... method is preferably used if the population is infinite:

a) Chit method b) Roulette method

c) Mechanical randomization d) Lottery method

14. The factor(s) which should not be kept in mind while selecting a sample size is/are:

    a) Lowest sampling error

    b) Good representative of population

    c) Must serve the purpose of investigation

    d) Highest sampling error

15. What will be the probability of selection of every unit in a population of size (N=3), if we have to draw a sample of size (n=2)?

    a) $\frac{1}{6}$ b) $\frac{1}{3}$

    c) $\frac{1}{9}$ d) $\frac{1}{4}$

16. The statistical consideration(s) not to be taken care of while doing the sample selection is/are:

    a) Degree of variability b) Desired confidence interval

    c) Level of precision d) Statistical regularity

17. The simplest type of probability sampling is:

    a) Simple random sampling b) Cluster sampling

    c) Systematic sampling d) Stratified sampling

18. The term used for proportion of risk a researcher is willing to take while dealing with data is:

    a) Standard error b) Standard deviation

    c) Margin of error d) Sampling error

19. As per the central limit theorem $z = \frac{x - \mu}{\sigma}$:

    a) Positively skewed variate b) Standard normal variate

    c) Negatively skewed variate d) Sample variate

20. The permissible level of confidence in medical science research is:

    a) 95% b) 99%

    c) 90% d) 100%

21. An increase in sample size will lead to .... in sampling error.
    a) Increase b) Decrease
    c) Constant d) Unity
22. An increase in confidence level will leads to .... in the confidence interval.
    a) Increase b) Decrease
    c) Constant d) Unity
23. The probability method has high efficiency and the smallest sample size in:
    a) Cluster sampling b) Simple random sampling
    c) Systematic sampling d) Stratified sampling
24. The numeric value of the Z score for 95 per cent confidence level is:
    a) 1.29 b) 1.96
    c) 2.54 d) 2.35
25. Calculate the sample size at a 95 per cent of confidence level with ± 5 per cent precision for evaluation of the adoption rate of high-density apple plantations in Himachal Pradesh.
    a) 280 b) 305
    c) 400 d) 385
26. Cochran's formula for estimation of sample size in simple random sampling for large samples:
    a) $\frac{(Z^2(pq)}{e^2}$ b) $\frac{Zp^2q}{e}$
    c) $\frac{Z(pq)^2}{e^2}$ d) $\frac{Z^2(pq)}{e}$
27. Cochran's formula for calculating ideal sample size in simple random sampling is not based on:
    a) Desired level of precision
    b) Desired level of confidence interval
    c) Estimated proportion of attributes present in the population
    d) Desired sampling error

28. Yamane's formula for calculating sample size is:

a) $\dfrac{N}{1+N(e)^2}$ b) $\dfrac{N^2}{1+N(e)^{0.5}}$

c) $\dfrac{N}{1\text{-}N(e)}$ a) $\dfrac{N}{1\text{-}N(e)^2}$

Key to MCQ's

1.	(b)	2.	(c)	3.	(c)	4.	(b)	5.	(b)
6.	(b)	7.	(a)	8.	(a)	9.	(b)	10.	(c)
11.	(c)	12.	(b)	13.	(c)	14.	(d)	15.	(c)
16.	(d)	17.	(a)	18.	(c)	19.	(b)	20.	(b)
21.	(b)	22.	(b)	23.	(b)	24.	(b)	25.	(d)
26.	(a)	27.	(d)	28.	(a)				

Let us do

1. Explain in detail Simple Random Sampling (SRS) with the help of suitable examples used in agricultural sciences.
2. Differentiate between the SRSWR and SRSWOR.
3. Explain the various methods to reduce biasness in selecting a random sample.
4. Prove that the sample mean is an unbiased estimator of population mean in SRSWR and SRSWOR.
5. What are the various factors which are responsible for determining the sample size?
6. Explain Cochran's and Yamane's formula for determining the sample size with the help of suitable examples.

4

Stratified Sampling

Stratified Sampling (Meaning and Concept)

The word 'stratify' comes from the Latin word, meaning 'to make layers', thus in stratified sampling the whole population is considered to be classified into some mutually exclusive layers or subgroups or classes also called strata. The strata do not overlap and they constitute the whole population.

Why the population is considered to be divided into different strata (or layers or subgroups or classes)?

If a population is not homogeneous in nature, the simple random sampling does not give a proper representation of the population. That's why if the population is heterogeneous, stratified random sampling is used, in which the heterogeneous population is divided into several layers also called subgroups or classes or strata that are more homogeneous than the total population. It means, we have to do stratification of the heterogeneous population. Stratification means splitting or division into layers or classes.

Stratified Sampling (Definition): A heterogeneous population of size N is divided into k relatively homogeneous subgroups or strata of size. N_1, N_2, ..., N_K such that $N = \sum_{i=1}^{k} N_i$ If a simple random sample (generally WOR) of size n_i (i = 1, 2, . . . , k) is drawn from each of the strata respectively, such that $n = \sum_{i=1}^{k} n_i$ the sample is termed a Stratified Random Sample of size n and the technique of drawing such a sample is known as Stratified Random Sampling.

Procedure of Drawing Stratified Sample

The following three points may be highly relevant in the context of stratified random sampling.

1. Formation of Strata.
2. Selection of Samples from each stratum.
3. Allocation of sample size to each stratum.

Stratified Random Sampling

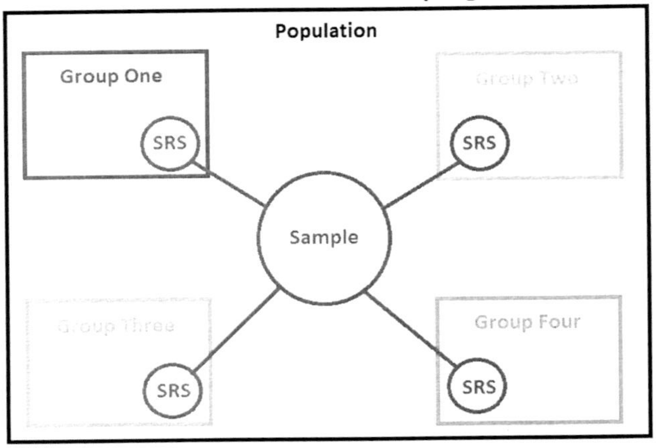

From this above diagram, we can easily understand these three main points which are required to draw a sample by using a stratified random sampling scheme. The whole square with black colour boundaries is considered to be the population. This population is heterogeneous in nature and has various different units. So we have to make strata or groups in such a way that units within each stratum are homogeneous in nature. In this figure there are four groups, group one marked with red colour, group two marked with yellow colour, group three marked with green colour and group four marked with blue colour. All these groups represent different colours because each group comprises with same units (homogeneous units) and comprises heterogeneous units between groups. Now from each group, we have to draw samples by using any sampling scheme according to the preference of the researcher. But, generally simple random sampling scheme is preferred and in our case, we also draw a sample with a simple random sampling procedure.

Since a stratified sample consists of units selected separately from each stratum, such a sample is expected to be a better representative of the population than a simple random sample selected from the whole population consisting of the heterogeneous units (with respect to character under study).

Example

Survey of manufacturing industries: Some factories may be very large. It may have huge fixed capital investment, a large number of workers, developing sophisticated appliances; while others may be of medium size, even tiny in nature.

Example

Socio Economic Survey: People may live in rural areas, urban localities, ordinary domestic houses, hostels, hospitals, jails etc.

Example

Full Concept: Suppose that a medical research scholar wants to determine the effect of a low-fat, high-carbohydrate diet on college students. He believes that the diet may have a different effect on students of different ages and genders so the population of students is divided into the following sub-population.

N_1: women of age 20 years or under

N_2: women over the age of 20 years

N_3: men of age 20 years or under

N_4: men over the age of 20 years

Where, $N = N_1 + N_2 + N_3 + N_4$

A simple random sample from each of these four non-overlapping sub-populations (strata) is drawn to form a stratified random sample of college students. This sample will we a representative sample for the said experiment on the diet.

Note: depending on the nature of the strata, different sampling designs can be used in different strata.

For example: In the absence of suitable size information, SRS can be used in a few strata, whereas probability proportional to size sampling can be used in the remaining strata when size information is available in those strata.

Main problems in stratification

1. The strata should not be overlapping and should together comprise the whole population.
2. With respect to character under study:
   - The strata should be homogeneous within the same strata/units.
   - The strata should be heterogeneous between other strata units.
3. If a researcher or investigator is facing some difficulties in stratifying a heterogeneous population concerning the characteristics under study, he/she should consider the administrative requirement as the basis for stratification.
4. If the limit of precision is given for a certain sub-population, it should be treated as a stratum.

Notations and Terminologies

We shall assume that the heterogeneous population of size N is divided into k different strata and that sampling within each stratum is conducted using SRSWOR technique.

1.	N		Population size
2.	N_i		Size of ith stratum in the population
3.	k		No. of strata
4.	Then N	$\sum_{i=1}^{k} N_i$	Total population
5.	Y_{ij}		Value of character under study for j^{th} unit of ith stratum
6.	$\bar{Y}_i$	$\frac{1}{N_i}\sum_{j=1}^{N_i} Y_{ij}$	Population mean of ith stratum
7.	$\bar{Y}_N$	$\frac{1}{N_i}\sum_{i=1}^{k}\sum_{j=1}^{N_i} Y_{ij}$	Overall population mean
		$\frac{1}{N}\sum_{j=1}^{N_i} N_i\bar{Y}_i = \sum_{j=1}^{N_i} W_i\bar{Y}_i$	
	Where $W_i = \frac{N_i}{N}$ are the population weights or population proportions for the ith stratum		
8.	S_i^2	$\frac{1}{N_i - 1}\sum_{i=1}^{N_i}(Y_{ij} - \bar{Y}_i)^2$	Mean square error of the ith stratum
9.	n		Sample size
10.	n_i		Size of the ith stratum
11.	Then n	$\sum_{i=1}^{k} n_i$	
12.	y_{ij}		Value of jth sampled unit from the ith stratum
13.	$\bar{y}_i$	$\frac{1}{n_i}\sum_{j=1}^{n_i} y_{ij}$	Sample mean of ith stratum
14.	f_i	$\frac{n_i}{N_i}$	Sampling fraction for i^{th} stratum
15.	s_i^2	$\frac{1}{n_i - 1}\sum_{j=1}^{n_i}(y_{ij} - \bar{y}_i)^2$	Sample mean square error of the ith stratum

16.	$\bar{y}_n$	$\frac{1}{n}\sum_{i=1}^{k}\sum_{j=1}^{N_i} y_{ij} = \frac{1}{n}\sum_{i=1}^{k} n_i \bar{y}_i$	Overall sample mean
17	$\bar{y}_{st}$	$\frac{1}{N}\sum_{i=1}^{k} N_i \bar{y}_i = \sum_{i=1}^{k} W_i \bar{y}_i$	

Where, $\bar{y}_{st}$ is the weighted mean of the strata sample means, the weights being equal to strata. It is important to note that $\bar{y}_n$ and $\bar{y}_{st}$ will be identical if $n \propto N$

In stratified sampling, the allocation of the sample to different strata is done by considering the following factors:

1. Stratum size i.e., the total number of units in each sample.
2. The variability within the stratum.
3. The cost associated with obtaining an observation from each stratum.

A good allocation of samples is one when maximum precision (or minimum variation) is obtained with minimum resources.

In other words, the criteria for allocation are to minimize the cost for a given variation or minimize the variation for a fixed cost, thus making the most effective use of the available resources.

Method of allocation of sample size

Under SRS of the stratified sample, let us consider the formula for the variance of $\bar{y}_{st}$

$$V(\bar{y}_{st}) = \frac{1}{N^2}\sum_{i=1}^{k} N_i (N_i - n_i)\frac{S_i^2}{n_i}$$

This equation depends on n_i, the number of units selected at random from the ith stratum.

Hence, the problem arises, what optimum value of n_i ($i = 1, 2, 3, \ldots, k$) can be chosen out of n, so that the variance is as small as possible.

For achieving these purposes, we consider four types of allocation namely:

1. Equal allocation
2. Proportional allocation
3. Neyman's allocation
4. Optimum allocation

1. Equal allocation: In this allocation method, the total sample size n is divided equally among all the strata so that size of each stratum is:

$$n_i = \frac{n}{k}$$

Where, K= 1, 2, 3,…., K

Total sample size for fixed cost is:

$$n = \frac{K(C\text{-}c_o)}{\sum_{i=1}^{k} c_i}$$

Unbiased estimator of population mean is:

$$\overline{Y}_{st} = \frac{1}{N}\sum_{i=1}^{n} N_i \overline{Y}_i$$

Variance of unbiased estimator of population mean is:

$$v(\overline{Y}_{st}) = \sum_{i=1}^{k} \frac{W_i^2(N_i\text{-}n_i)}{N_i n_i} S_i^2$$

Limits of confidence interval:

$$\overline{y}_{st} \pm 2\sqrt{v(\overline{y}_{st})}$$

2. Proportional allocation: This allocation was originally proposed by Bowley in 1926. Under this allocation, the sub-sample size from each stratum is proportional to the size of the sub-population in the stratum.

$$n_i \propto N_i \text{ for } i = 1, 2, 3, \dots, k$$

$$n_i = \frac{n}{N} N_i$$

$$n = \frac{(C\text{-}c_o)}{\Sigma_{i=1}^{k} W_i c_i}$$

Population mean of unbiased estimator is:

$$\overline{y}_{st} = \frac{1}{N} \Sigma_{i=1}^{n} N_i \overline{y}_i$$

The estimator of variance is:

$$v(\overline{y}_{st}) = \sum_{i=1}^{k} \frac{W_i^2(N_i\text{-}n_i)}{N_i n_i} s_i^2$$

Limits of confidence interval is:

$$\overline{y}_{st} \pm 2\sqrt{v(\overline{y}_{st})}$$

Main drawback of proportional allocation

Proportional allocations do not take into account any factor other than strata size. They completely ignore the internal structure of strata like within stratum variability etc. and hence, it is desired to consider an allocation scheme which takes into account these aspects.

3. Optimum allocation: The optimum allocation method is based on the cost aspect of the survey.

Cost Function: let C_t be the total cost of the survey. The cost is generally measured in terms of expenditure incurred on the survey.

Total sample size:

$n = \sum_{i=1}^{k} n_i$

$$\text{where, } n_i = \frac{(C - c_o)\dfrac{(W_i S_i)}{\sqrt{c_i}}}{\sum_{i=1}^{k} w_i s_i \sqrt{c_i}}$$

Unbiased estimator of population mean is:

$$\overline{y}_{st} = \frac{1}{N}\sum_{i=1}^{k} N_i \overline{y}_i$$

Variance of unbiased estimator of population mean is:

$$v(\overline{y}_{st}) = \sum_{i=1}^{k} \frac{W_i^2 (N_i - n_i)}{N_i n_i} S_i^2$$

Limits of confidence interval is:

$$\overline{y}_{st} \pm 2\sqrt{v(\overline{y}_{st})}$$

4. Neyman's allocation: This allocation is based on the following assumptions:

i. The sampling cost per unit among different strata is same.

ii. The size of the sample is fixed.

The sample sizes are allocated by considering the size of strata as well as variability i.e.,

$n_i \propto N_i S_i$

$$n_i = \frac{W_i S_i}{\Sigma_{i=1}^{k} W_i S_i} \text{ or } n_i = \frac{N_i S_i}{\Sigma_{i=1}^{k} N_i S_i}$$

Total cost of sample size for fixed cost is: In this particular kind of allocation cost per element is just the same for all the strata i.e.,

$c_i = c'$ for each i

$C = c_o + c' n$

and $n = \frac{C\text{-}c_o}{c^i}$

Unbiased estimator of population mean is:

$\bar{y}_{st} = \frac{1}{N} \sum_{i=1}^{k} N_i \bar{y}_i$

Variance of unbiased estimator of population mean is:

$v(\bar{y}_{st}) = \sum_{i=1}^{k} \frac{W_i^2 (N_i - n_i)}{N_i n_i} s_i^2$

Limits of confidence interval is:

$\bar{y}_{st} \pm 2\sqrt{v(\bar{y}_{st})}$

Practical example

A survey was conducted at the university of agricultural sciences to know the average income of staff. For this, the income is divided into three groups such as low, medium and high and these divisions are known as strata. Because the whole staff working in the university has different incomes so that the population from where the sample is to be selected is a heterogeneous group. Due to this heterogeneity present in the whole population, we divided the population into different groups. In this situation with the presence of a heterogeneous group, stratified random sampling is used to draw the sample for the study.

Let us assume that the total population of working staff in the university is 5000 and this population size is divided into three groups based on income. The size and standard deviation of these three groups are given below:

Stratum	Low income	Medium income	High income
Size	1800	2000	1200
SD	12	10	8

From this population, a sample of size 350 is to be drawn. Determining the size of the sample from these three groups is done by proportional allocation and optimum allocation.

In such type of research or survey, we can easily get a proper representative sample which satisfies the whole population.

Low income = N_1=1800

Medium income = N_2 = 2000

High income = N_3 = 1200

Whole population = N = 5000

Sample to be drawn n = 350

Now, it is important that from each group how you select the sample i.e. from proportional allocation.

Sample size to be drawn from low income i.e. $n_1 = N_1 \times \frac{n}{N} = 1800 \times \frac{350}{5000} = 126$

$$\therefore n_2 = N_2 \times \frac{n}{N} = 2000 \times \frac{350}{5000} = 140$$

$$\therefore n_3 = N_3 \times \frac{n}{N} = 1200 \times \frac{350}{5000} = 84$$

Now, the sample is drawn from the optimum allocation.

In that case, SD is given i.e. $s_1 = 12$, $s_2 = 10$, $s_3 = 8$

$$\therefore n_1 = n \times \frac{N_1 s_1}{\sum N_i s_i} = 350 \times \frac{21600}{51200} = 147.65 \approx 148$$

$$\therefore n_2 = n \times \frac{N_2 s_2}{\sum N_i s_i} = 350 \times \frac{20000}{51200} = 136.71 \approx 137$$

$$\therefore n_3 = n \times \frac{N_3 s_3}{\sum N_i s_i} = 350 \times \frac{9600}{51200} = 65.62 \approx 66$$

Multiple Choice Questions (MCQ's)

1. The word 'stratify' is extracted from..........language:

   a) Latin b) Greek

   c) Roman d) Italian

2. The stratified random is a type of:

   a) Non-probability sampling b) Probability Sampling

   c) Stratified sampling d) Systematic sampling

3. If the objective is to minimize the survey cost which method of sample allocation is most appropriate:

   a) Optimum b) Neyman

   c) Equal d) Proportional

4. If the population given in the question is heterogeneous, which method of sampling is most suitable?

   a) Simple random sampling b) Stratified random sampling

   c) Mixed sampling d) Systematic sampling

5. The groups into which the universe is divided in case of stratified random sampling are called as:

   a) Strata b) Clusters

   c) Sub groups d) Units

6. The strata so formed after stratification in stratified sampling are.......... in nature:

   a) Heterogeneous b) Mixed

   c) Homogeneous d) Diverse

7. If WHO surveys to know the impact of co-vaccine on different age groups in India, which sampling technique is most suitable?

   a) Cluster sampling b) Systematic sampling

   c) Stratified random sampling d) Simple random sampling

8. Which of these is an incorrect statement about stratified random sampling?

   a) Each unit in the universe can be assigned to a single stratum

   b) Stratum should be non-overlapping

   c) The probability of selection of each and every unit in the population is same

   d) Each unit in the universe can not be assigned to a single stratum

9. The sample size allocated under stratified random sampling, via which of the following techniques?

   a) Equal b) Neyman

   c) Optimum d) All of these

10. Which of the following allocation technique is used to divide the sample size equally among different strata:

   a) Optimum b) Neyman

   c) Equal d) Proportional

11. Which of these are not the major advantages of stratified random sampling over other types of sampling techniques?

   a) Lower cost of conducting survey

   b) Administrative convenience

   c) Minimal chance to exclude any important component

   d) Higher cost of conducting survey

12. Which factor(s) are not to be kept in mind when stratifying the given population in stratified random sampling?

   a) Stratum size

   b) Cost involved in taking observation from each stratum

   c) Variability within strata

   d) Homogeneity within strata

13. The formula for stratum size in equal allocation is:

   a) $n_i = \frac{n}{k}$ b) $n_i = \frac{k}{n}$

   c) $n_i = nk$ d) $n_i = \frac{1}{k}$

14. The proportional allocation was initially proposed by:
    a) Hawkins b) Bowley
    c) Yates d) Neyman
15. The allocation of strata involving a selection of sub-sample size proportional to population size in each stratum is:
    a) Neyman allocation b) Proportional allocation
    c) Optimum allocation d) Equal allocation
16. The type of allocation whereby the sample size allocation is based on both stratum size and stratum variability:
    a) Equal allocation b) Optimum allocation
    c) Proportional allocation d) Neyman allocation
17. ....is the major consideration while allocating sample size under optimum allocation.
    a) Stratum size
    b) Stratum variance
    c) Cost of conducting a survey
    d) Cost of conducting stratum mean
18. In.......................type of allocation stratum size is the sole criteria of selection of sample:
    a) Equal allocation b) Optimum allocation
    c) Proportional allocation d) Neyman allocation

Key to MCQ's

1.	(a)	2.	(b)	3.	(a)	4.	(b)	5.	(a)
6.	(c)	7.	(c)	8.	(a)	9.	(d)	10.	(c)
11.	(d)	12.	(d)	13.	(a)	14.	(b)	15.	(b)
16.	(d)	17.	(c)	18.	(c)				

Let us do

1. A researcher is given the problem to estimate the average time devoted every week to physical activities by the sportsmen at a sports academy. The sportsmen in the academy belonged to three categories i.e., under 19, from 19 to 30 group and above 30 group. The numbers of persons

in each group were 900, 660 and 500 respectively. The researcher made three strata namely; under 19, 19-30 and above 30. He selected random samples of 20, 15 and 10 from three strata respectively to make a total of 45.

Time utilized weekly for sports activities (in Hrs)

Strata I (20)				Strata II (15)			Strata III (10)	
5	12	8	9	9	10	10	6	8
9	3	7	10	12	9	9	9	87
6	6	8	10	12	11	10	8	6
12	10	7	12	13	10	6	7	5
13	11	9	15	14	12	7	9	

Estimate the average time devoted per week by sportsmen towards physical activities in the sports academy. Also, give variability in time devoted as well as build up the confidence interval for this average.

2. Briefly discuss stratified random sampling. Justify the situations as where stratified random sampling is used and why?
3. Stratifying a population is one of the most important task in stratified random sampling. Give some important technical points to be taken care of while stratifying the population.
4. Enlist various sample allocation methods in stratified sampling. Also, give the merits and demerits of each sampling method.
5. A stratified sample of 100 is to be selected from a heterogeneous population of 5000 units, sub-divided into three strata of stratum sizes; 2100, 1500 and 1400 units. What will be the sample size in each of the cases in equal, proportional, Neyman and optimum allocation?
6. Express your views about using stratified random sampling in each of the following cases:
    a) Per head mutton intake of mammals and reptiles group at a zoo.
    b) Proportion of scheduled tribes in a district.
    c) Average height and weight of Montessori boys and girls in a school.
7. Investigator has to evaluate the sociopolitical status of employees at a university having 220 class-I employees, 300 class-II, 150 class-III and 400 class-IV. For this, he took samples of 22, 30, 15 and 40 employees from Class I, II, III and IV employees respectively. Is this a proportionate allocation? Discuss why and why not?

8. What are the major challenges an investigator is confronted with while stratifying a population?

9. State the method of sample allocation most suitable under the following circumstances:

   a) Total cost of a survey is fixed and variance minimization is the sole objective.

   b) Objective is to minimize the survey cost.

5

Systematic Sampling

Systematic Sampling (Meaning and Concept)

In this chapter, we are going to discuss the systematic random sampling technique. This is a random / probability / quantitative sampling method. In our previous chapter, we discussed the stratified random sampling technique. We have said that this technique is used when the population is heterogeneous and needs to be stratified into homogeneous groups. We have also discussed steps to be followed when using stratified random sampling as well as the strengths and limitations of stratified random sampling. Systematic sampling is a commonly employed technique if a complete and up-to-date list of the sampling units is available. The procedure involves choosing the very first object arbitrarily and the remaining members of the sample will be selected naturally according to any pre-decided sequence containing regular spacing of units. In this method, units of the population are numerically, geographically, and alphabetically arranged either in ascending or descending order (generally ascending order is preferred). When the ordering of the elements is related to the characteristic of interest, systematic sampling increases the representativeness of the sample. Every n^{th} item or unit of the numbered items is selected as a sample item. This periodic interval is referred to as "***sampling interval***" and this method is known as "***quasi-random sampling***". This sampling is used when units are uniformly distributed and there is no fixed pattern and the population is homogeneous.

In what conditions systematic sampling is used

1. Complete sampling frame is required.
2. Precision level of the study is high.
3. If we have information on all the population units.

Practical application

1. Systematic sampling is quite preferably applicable in forest surveys to estimate the volume of trees.

2. It is most commonly used in milk yield surveys for estimating the lactating yield.

Steps in doing a systematic sampling

1. Define our target population.
2. Figure out our ideal sample size.
3. Assign a number to every unit of the sample.
4. Determination of sampling interval.
5. Between 1 and the sampling interval, an arbitrary number is chosen.
6. Repeat the sampling interval to choose your subsequent elements.

Merits

1. Very simple method and sample are easily determined and analyzed.
2. Easy to use in comparison to SRS for the large population.
3. There is hardly any possibility of personal bias in this method.
4. Less chance of manipulation of data by the researcher.
5. Cost-effective.
6. In sampling techniques that involve the selection of clusters such as cluster sampling, stratified sampling and convenience sampling (non-probability sampling) there are some chances on the part of the researcher/statistician, which are absent in this sampling as the units are at a specified interval from each other.
7. Minimal risk factor is involved as the systematic samples are well distributed over the population and there is no risk of any contiguous part of it left unrepresented.
8. It is also efficient for a population with autocorrelations.
9. This method offers an advantage of even arrangement of objects in the sample, the population given is diversified.

Note: Systematic sampling should, however, be applied with substantial care in case of periodicity is present in the population. Where the values of study variables for the units given in the frame tend to increase and decrease and then again increase in a cyclic manner over a fixed period. For, a periodic population if the sampling interval is an odd multiple of half of the period of the cycle, a systematic sampling estimator provides the estimator of the mean with appreciable efficiency. However, if the sampling interval is an integral multiple of the period of the cycle, simple random sampling is the better option over systematic sampling.

Demerits

1. Every item in the population does not get an equal chance of being selected because only the first item is selected on the basis of random sampling.
2. The method fails in the case of a homogeneous population.
3. It can produce skewed results if there exists some pattern.
4. Less randomness.
5. Estimation of the sampling variance of estimators with a single sample may not provide reliable results.

Systematic Sampling is of two types

1. Linear systematic sampling
2. Circular systematic sampling

Linear Systematic Sampling

The most widely used procedure of systematic sampling is linear systematic sampling. Let us assume that in order to choose a sample of size n from a population comprising of N units. The N units of the population are encompassed from 1 to N in some order and a sample of size n is to be chosen.

Case-I: If $N = nk$ so, $k = N/n$

Where, k is an integer also known as sampling interval.

The systematic sampling embraces of drawing a random number, suppose the random number to be drawn in the first instance such that $i \leq k$, and select every k^{th} unit corresponding to this number eventually. In consequence, the systematic sample of size n will comprise of the units $i, i+k, i+2k, \ldots, i+(n-1)k$, where, i is the random start. Such a procedure is termed as Linear systematic sampling.

From the above figure, N = 20, we want to draw a sample of size n = 5, then k = N/n = 4. Therefore, we want to select a random number from 1 to 4, suppose 1 is selected at random then the sample consists 1, 5, 9, 13, 17 and if random numbers 2, 3 and 4 are selected then, these three possible samples contain units as 2, 6, 10, 14, 18 for random number selected as 3 then samples to be selected as 3, 7, 11, 15, 19 similarly if the random number is 4 then selected samples as 4, 8, 12, 16, 20 respectively.

Example

Suppose we have a small forest consisting of a total of 100 trees (N) numbered in an ordered manner, out of which a sample (n) of 5 trees is to be drawn. What will be the sampling units to be drawn?

Solution

If N = 100, n = 5 therefore, k = 100/5 = 20. Let the random number chosen from 1 to 20 be 16, with 16 as a random start, the units bearing serial numbers 16, 36, 56, 76 and 96 will be selected in the sample.

Circular Systematic Sampling

Circular Systematic Sampling allows us to create samples, we have to select our sample subjects from an interval in a systematic random sampling manner. In circular systematic sampling, our selection will restart from the start point once we have considered the entire population.

Case-II: If $N \neq nk$

Let us suppose N = 20, and we want to draw a sample of size n = 7, then k = N/n = 2.86 $\approx$ 3 therefore if we have to select a random number from 1 to 3, suppose, random number 1 is selected then the sample comprises of 1, 4, 7,

10, 13, 16, 19. If we select 2 as a random number then our sample will contain 2, 5, 8, 11, 14, 17 and 20. But, if the number selected is 3 then only six units will be selected i.e., 3, 6, 9, 12, 15, 18 as we cannot select 7th unit which is 21 as it is not present in the population. As a consequence, the actual sample size may be different from the required one. In such situations, the sample mean also does not remain the unbiased estimate of the population mean. These shortcomings can be overcome by using a sampling procedure known to be circular systematic sampling.

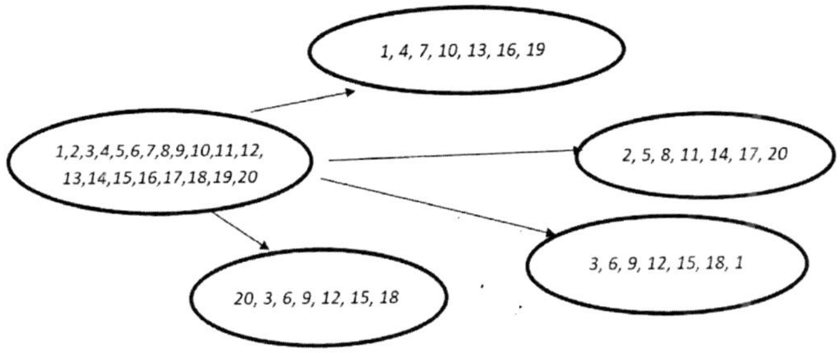

This type of sampling was first used by D B Lahiri (1952). Here k is taken as the nearest integer of N/n.

- Select a random number from 1 to N.
- Let it be m, then select every $(m + jk)^{th}$ unit when, $m + jk < N$ then select every $(m + jk\text{-}N)^{th}$ unit. When, $m + jk > N$ putting $j = 1, 2, 3, \ldots$ till n units are selected. Therefore in the above figure, our two samples were chosen very easily but for the third sample where our first random unit was 3, the last unit was selected by using the formula $(m + jk - N)^{th}$ so the 1st unit will be selected. Suppose for 4th sample let random number 20 be selected then the second unit selected will be 3, not 23 as this number is not present in the population, and we continue this procedure until the desired sample size is obtained.

Merits

1. Provides constant sample size.
2. Sample mean remains an unbiased estimate of the population mean.

Demerits

In this type, it is not possible to obtain an unbiased estimate of the sample variance of the estimator from a single systematic sample.

Points of distinction between linear and circular systematic sampling

Criteria	Linear Systematic Sampling	Circular Systematic Sampling
Sample selection	Creation of sample depends upon sampling interval (k)	Sample is created from the entire population
Start and end points	The numbers selected at start and end points are different	Entire population is considered for selection of units once the starting point is chosen
Ordering of sample units	Linear ordering of units before selection	Circular ordering of units is there

Some important formulae for estimating mean and total:

- **Estimator of population mean is:**

$$\bar{y}_{sys} = \frac{1}{n}\Sigma_i^n y_i$$

- **Variance of the estimator $\bar{y}_{sys}$:**

$$V(\bar{y}_{sys}) = \frac{1}{k}\Sigma_{r=1}^{k}(\bar{y}_{sys} - \bar{Y})_r^2 \text{ (For linear systematic sampling)}$$

$$V(\bar{y}_{sys}) = \frac{1}{N}\sum_1^N(\bar{y}_{sys} - \bar{Y})^2_r \text{ (For circular systematic sampling)}$$

Where, $(\bar{y}_{sys} - \bar{Y})_r$ is the difference between the systematic sample mean corresponding to random start r and the population mean $\bar{Y}$.

- **Estimator of variance is:**

$$v = (\bar{y}_{sys}) \frac{N-n}{2Nn\,(n-1)}\sum_{i=1}^{n-1}(y_{i+1}-y_i)^2$$

$$v\,(\bar{y}_{sys}) = \frac{N-n}{Nn\,(n-1)}\sum_{i=1}^{n}(y_i - \bar{y}_{sys})^2 \text{ (for random population)}$$

**Source: Singh and Mangat, 1996*

Example

A forest land consists of 4000 deodar trees. The Forest department attempted to estimate the timber volume as they reached their rotation age (120 years). 1 in 100 systematic samples is chosen. Estimate the total timber volume and confidence interval for it. The database on observed timber volume for selected trees is given below.

Serial No.	Timber Vol.	Serial No.	Timber Vol.	Serial No.	Timber Vol.	Serial No.	Timber Vol.	Serial No.	Timber Vol.	Serial No.	Timber Vol.
50	2.72	750	3.20	1450	2.91	2150	3.60	2850	2.51	3550	3.1
150	2.29	850	2.03	1550	2.66	2250	3.64	2950	1.26	3650	3.50
250	3.08	950	3.17	1650	2.56	2350	2.43	3050	1.56	3750	3.0
350	2.29	1050	3.47	1750	3.26	2450	2.89	3150	3.00	3850	3.14
450	2.07	1150	4.86	1850	4.49	2550	3.63	3250	2.94	3950	2.56
550	4.77	1250	2.17	1950	2.26	2650	2.23	3350	3.90		
650	1.63	1350	1.63	2050	2.31	2750	2.40	3450	1.60		

Solution

Here, N = 4000 and k = 100

Therefore, if r is assumed to be the first random number chosen and let it be 50. Therefore, the deodar trees bearing serial numbers, 50, 150, 250, …, 3950 will be included in the sample.

$\bar{Y}_{SYS} = N\bar{y}_{sys} = N/n \sum_{i=1}^{n} y_i$

$= 4000/40\ (2.72+2.29+\ldots 2.56)$

$=100\times 112.72$

$=11272$

Now $V(\bar{Y}_{sys}) = N^2 v(\bar{y}_{sys}) = \frac{N(N-n)}{2n(n-1)} \sum_{i=1}^{n-1} (y_{i+1} - y_i)^2$

$= \frac{N(N-n)}{2n(n-1)} [(y_2 - y_1)^2 + \ldots + (y_{40} - y_{39})^2]$

$= \frac{4000(4000-40)}{2(40)(39)} [(2.29\text{-}2.72)^2 + \ldots + (2.56\text{-}3.14)^2)]$

$= 5076.92 \times (60.85)$

$= 308934.8$

Confidence Interval $= N\,\bar{y}_{sys} \pm 2N\sqrt{v(\bar{y}_{sys})}$

$= \hat{Y}_{SYS} \pm 2\sqrt{v(\hat{Y}_{sys})}$

$=11272 \pm 2\sqrt{308934.8}$

$=11271 \pm 2\times 555.82$

$=11272 \pm 1111.64$

$=10159.36$ and 12383.36

Example

On a very chilling cold day, 220 rural women had gone into a forest to collect dry wooden logs. It was initially aimed to estimate the total dry wood collected at the day end. But the idea was later dropped as estimation of total wood weight by all 220 ladies is a cumbersome task. Therefore, it was decided to weigh wood from only 21 ladies selected via circular systematic sampling. Discuss the procedure for the selection of the required sample to estimate the total wood collected based on wood collected by sample of 21 women selected.

Table: Dry wood collected (In quintals) by 21 ladies.

S.No.	Wood weight	S.No.	Wood weight	S.No.	Wood weight
80	3.96	7	3.92	154	2.20
101	4.83	28	5.11	175	8.0
122	5.96	49	8.36	196	6.66
143	10.21	70	9.20	217	7.20
164	8.30	91	8.80	18	8.80
185	5.12	112	2.96	39	7.70
206	4.76	133	1.20	60	8.00

Solution

Here, N = 220 and n = 21

Since N/n =220/21 =10.48 which is not a whole number. Therefore, the value of the sampling interval is taken as 11 and circular systematic sampling is used for selecting the women.

Therefore, let "r = 80" be the random number selected between 1 to 220. So, the women with serial numbers 80, 101, …, 39, 60 will be selected.

Mean

$$\bar{Y}_{sys} = \bar{y}_{sys} N = N/n \sum_{i=1}^{n} y_i$$

$$= 220/21\ (3.96 + 4.83 + \ldots + 9)$$

$$= (220)\ (131.25)/21$$

$$= 1375$$

Estimated Variance

$$V(\hat{Y}_{sys}) = N^2\, V(\bar{y}_{sys}) = \frac{N(N-n)}{2n(n-1)} \Sigma_{(i=1)}^{(n-1)}\ (y_{i+1} - y_i)^2$$

$$\frac{220(220-21)}{2(21)(20)} [(4.83\text{-}3.96)^2 + \ldots (8\text{-}7.7)^2]$$

$$= 52.12 \times 125.33$$

$$= 6531.90$$

Confidence Interval

$$\hat{Y}_{sys} \pm 2\sqrt{v(\bar{Y}_{sys})}$$

$$= 1375\ \sqrt{2\sqrt{6531.90}}$$

$$=1213.36 \text{ and } 1536.64$$

Multiple Choice Questions (MCQ's)

1. When the population is heterogeneous and needs to be stratified into homogeneous groups in that condition which sampling method is preferable.

   a) Cluster sampling   b) Quota sampling

   c) Systematic sampling   d) Purposive sampling

2. In............... type of sampling method, the units of the population are numerically, geographically, and alphabetically arranged either in ascending or descending order.

   a) Systematic sampling   b) Stratified sampling

   c) Cluster sampling   d) All the above.

3. In what conditions systematic sampling is used.

   a) Complete sampling frame is required.

   b) Precision level of the study is high.

   c) If we have information of all the population units.

   d) In all the above conditions.

4. Which type of sampling scheme is quite preferable in forest surveys to estimate the volume of trees.

   a) Stratified sampling   b) Systematic sampling

   c) Cluster sampling   d) Quota sampling

5. Which type of sampling scheme is most commonly used in milk yield surveys for estimating the lactating yield.

   a) Purposive sampling   b) Systematic sampling

   c) Stratified sampling   d) Cluster sampling

6. In .... sampling scheme the sampling interval is required.

   a) Systematic   b) Judgemental

   c) Quota   d) Cluster

7. Which type of sampling scheme is easy to use in comparison to SRS for the large population.

   a) Stratified sampling   b) Cluster sampling

   c) Quota sampling   d) Systematic sampling

8. If $N = nk$ which type of sampling scheme is used.

   a) Circular systematic sampling   b) Linear systematic sampling

   c) Systematic random sampling   d) None of the above

9. If $N \neq nk$ which type of sampling scheme is used.

   a) Circular systematic sampling   b) Linear systematic sampling

   c) Systematic random sampling   d) None of the above

10. If a population had 175 units and we want to take a sample of 35 units by using systematic sampling, what would be interval size be.

    a) 35   b) 175

    c) 5   d) 25

11. The agricultural department has 50 workers. In order to find out some information about the department as a whole, we want to take a sample of 10 workers to interview. If we have ordered list of workers numbered from 1 to 50, and we start at worker number 4, which workers would be included in our sample.

    a) 4, 9, 14, 19, 24, 29, 34, 39, 44, 49

    b) 5, 10, 15, 20, 25, 30, 35, 40, 45, 50

    c) 3, 8, 13, 18, 23, 28, 33, 38, 43, 48

    d) 4, 8, 12, 16, 20, 24, 28, 32, 36, 40

12. Which method is required for selecting samples based on a system of intervals in a numbered population?

    a) Simple random sampling   b) Systematic random sampling

    c) Cluster sampling   d) Law of large numbers

13. When we are doing systematic sampling, we must remember that all outcomes must be given .... chance of being selected.

    a) More than double   b) Less than half of the

    c) Equal   d) Twice the

14. To use systematic random sampling, we must first create a list of our population, and then.........................

    a) Select a beginning number   b) Create a list of employees

    c) Pull all names in a hat   d) Select an interval

15. In the systematic sampling, the value of k is classified as:
    a) Sub stage interval
    b) Sampling interval
    c) Multistage interval
    d) Secondary stage interval
16. The elements in the sample with specific characteristics are divided to sample sizes to calculate:
    a) Expected deviated proportion
    b) Expected mean proportion
    c) Population proportion
    d) Sample proportion

Key to MCQ's

1.	(c)	2.	(a)	3.	(d)	4.	(b)	5.	(b)
6.	(a)	7.	(d)	8.	(b)	9.	(a)	10.	(c)
11.	(a)	12.	(b)	13.	(c)	14.	(d)	15.	(b)
16.	(d)								

Let Us Do

1. Explain the meaning and concept of systematic sampling.
2. In what conditions was the systematic sampling used?
3. What are the various steps while we draw a sample with the help of systematic sampling?
4. Explain the merits and demerits of systematic sampling.
5. Explain the concept of linear systematic sampling with the help of an example.
6. Explain the concept of circular systematic sampling with the help of an example.
7. Differentiate between linear systematic sampling and circular systematic sampling.

6

Cluster Sampling

Cluster Sampling (Meaning and Concept)

Cluster sampling is a special type of sampling which is used when we have to make a large survey such as a survey for the whole country and whole state etc. to know the privilege. Under this method, the total population is divided into some recognizable subdivisions known as clusters. Out of all the clusters, a given number of clusters are chosen at random. All the items covered by the selected clusters are included in the sample.

A sampling technique in which we divide the whole population into groups or subgroups in such a way that there is heterogeneity within each group and homogeneity among the groups and then a simple random sample of n groups is selected from a total of N groups of population and finally all the units of selected clusters from our sample.

Why we use cluster sampling

i. When a sampling frame is not available then SRS, Systematic Sampling, Stratified Sampling etc. are not possible then cluster sampling is to be used.

ii. Even if a sampling frame is available but the population is large and units are scattered or dispersed, this will require a lot of travel expenditure to collect the data, then cluster sampling is easy to use and economical in such situations.

iii. When information is required for groups of units (clusters) instead of individual units.

It is a probability sampling technique that is commonly employed to study large populations that are geographically dispersed. Multiple clusters of people from the target population are created. The dispersal of the characteristics within the clusters so created must be comparable to those of the distribution of the universe.

Remark: Stratification is done intentionally and for administrative convenience

While using cluster sampling, the following points should be kept in mind

i. Define your target population.

ii. Divide your samples into clusters.

iii. The cluster should be small as possible, consistent with the cost and limitations of the survey.

iv. Clusters created should be accurate prototypes of the population.

v. Different clusters so formed should have alike properties.

vi. Clusters should be mutually homogeneous but internally heterogeneous.

vii. Clusters should be mutually exclusive.

viii. Randomly select clusters as our samples.

ix. Collect data from the chosen sample.

Advantages

i. Convenient and economical.

ii. The loss in efficiency is likely to be balanced by a reduction in cost.

iii. In many situations like widely spread geographical populations, only cluster sampling is applicable or best to use while other sampling techniques do not.

iv. The knowledge of a variety of areas is provided by this sampling technique.

v. Cluster sampling permits each accumulation of large samples.

vi. Cluster sampling may combine the advantages of both random sampling as well as stratified sampling.

vii. Selecting the clusters of units in a universe is more economical and rapid than observing the units randomly from it.

viii. It is useful when we don't have access to all the subjects of the population.

Disadvantages

i. This sampling technique is comparatively less efficient than a sampling of individual units.

ii. The efficiency of cluster sampling is likely to decrease with the increase in cluster size.

iii. This sampling technique ignores some portion of the population so the selected samples are likely to be not representative of the population.

iv. It can become more complex in planning.

v. The subsets so created in cluster sampling should not be dissimilar in size otherwise it may cause sampling bias.

vi. Sometimes, an insufficient number of cases to enhance the accuracy of the sample are not observed, which may lead to an overlapping effect.

vii. In many instances, the cluster so created doesn't have homogeneity from within, which may result in sampling errors.

viii. The results observed in cluster sampling in one locality may not be suitable for some other locality.

Examples

1. If we are interested in obtaining the income of opinion data in a city, the whole city may be divided into N different blocks or localities (which determine the clusters and a simple random sample of n blocks is drawn. The individuals in the selected blocks determine the cluster sample.

2. A survey was conducted to know the opinion of agricultural scientists across India about the various farm bills, 2020. Here, the population includes agricultural scientists across the country. Therefore if we go for simple random sampling it will turn out to be a clumsy process. Hence, to make the process convenient, easier and feasible we will prefer cluster sampling. Various agricultural universities in India will be our clusters and in the next step, the required number of clusters (agricultural universities) will be selected randomly and then each scientist from these selected clusters will be surveyed.

3. WHO recommended a modified form of cluster sampling called "30×7" for the evaluation of the vaccination drive. A complete list of cluster villages was prepared out of which 30 were observed through PPS sampling. Out of the cluster villages, 7 units were randomly selected as ultimate respondents.

Suppose we have N clusters in our population under investigation and each cluster has M elements, therefore total population units will be N×M. If we select n clusters out of N clusters in the population by applying various sampling procedures *viz*; simple random sampling, probability proportional to size sampling, systematic sampling to cluster sampling by treating clusters as sampling units.

Note: It is a superficial concept in reality cluster doesn't have equal elements. It's just done to develop theory.

Difference between cluster sampling and stratified sampling:

S.No.	Cluster sampling	Stratified sampling
1.	Groups formed are called clusters.	Groups formed are called strata.
2.	The whole group is selected as a sample.	A sample is a single item.
3.	Clusters are so constructed that they are internally heterogeneous and externally homogenous.	Strata are so constructed that they are internally homogenous and externally heterogeneous.
4.	The cluster itself is a sampling unit.	Elements within strata are sampling units.
5.	Clusters are naturally occurring groups.	Strata are imposed by the researcher.
6.	To reduce time and cost and improve efficiency.	This sampling is used to increase precision and representation.

Single-stage and Multistage cluster sampling

- In single-stage cluster sampling, the universe is first divided into appropriate clusters by simple random sampling, thereafter, each and every object within the selected clusters is observed.
- Cluster sampling can also be performed in many stages known as multistage cluster sampling. For instance, if a researcher is willing to know the average annual expenditure on agriculture machinery in a country, cluster sampling in the first stage observes various states where agriculture is the main occupation. In the second stage, selected districts from each state and in the third selected villages will be surveyed and farmers will be interviewed in the fourth stage.

Area sampling

- A Special type of cluster sampling is area sampling. Area sampling is mostly preferred and best suited when research is confined to a particular geographical area or location.
- For instance, the horticulture department in Himachal Pradesh wants to install a weather station for Weather Based Crop Insurance Scheme to check the validity of applications for claim settlements of policyholders. Then the stations will be installed in different locations where weather parameters vary like wind speed, Relative humidity and temperature etc.

Multiple Choice Questions (MCQ's)

1. Which type of sampling technique is used to make a large survey area?

    a) Simple random sampling b) Systematic sampling

    c) Stratified sampling d) Cluster sampling

2. The total population is divided into some recognizable subdivisions known as:

    a) Sample b) Strata

    c) Cluster d) Stratum

3. If there is heterogeneity within each group is known as:

    a) Cluster b) Strata

    c) Stratum d) Sample

4. If there is homogeneity among groups is known as:

    a) Cluster b) Strata

    c) Stratum d) Sample

5. When the sampling frame is not available then ....scheme is used.

    a) Multiphase sampling b) Cluster sampling

    c) Stratified sampling d) Systematic sampling

6. When a sampling frame is available but the population is large and units are scattered then ....scheme is used.

    a) Simple random sampling b) Systematic sampling

    c) Cluster sampling d) Stratified sampling

7. When we require information for various groups of units not by individual units.

    a) Cluster sampling b) Mixed sampling

    c) Stratified sampling d) Quota sampling

8. The size of the cluster should be ....as possible.

    a) Large b) Small

    c) Medium d) Constant

9. The cluster should be .... with the cost.
   a) Unbiased b) Consistent
   c) Sufficient d) Efficient
10. Clusters so created should be accurate .... of the population
    a) Samples b) Parameters
    c) Estimators d) Variables
11. The efficiency of cluster sampling is likely to be .... with the increase in cluster size.
    a) Increase b) Decrease
    c) Constant d) Proportional
12. The efficiency of cluster sampling is likely to be .... with the decrease in cluster size.
    a) Increase b) Decrease
    c) Constant d) Proportional
13. All the statements are true about the cluster sampling concept.
    a) Samples are similar to those in simple random sampling
    b) It is a rapid and simple method
    c) The sample size may vary according to the study design
    d) It is a type of probability sampling
14. In cluster sampling, the elements of selected clusters are classified as:
    a) Elementary units b) Primary units
    c) Secondary units d) Proportional units
15. A researcher intends to study the adjustment problems of students who are slum dwellers. What should be the sampling procedure used for such a study?
    a) Simple random sampling b) Stratified random sampling
    c) Systematic sampling d) Cluster sampling

16. ……………. is a sampling method where different groups within a population are used as a sample.

    a) Random sampling b) Simple random sampling

    c) Cluster sampling d) Systematic sampling

17. Which is a sampling method that involves separating the population into clusters, and then selecting random samples from each of the clusters.

    a) Two-way cluster sampling b) Simple random sampling

    c) Stratified sampling d) Systematic sampling

18. The term cluster was first used by ……………..

    a) Thompson (1943) b) Hurwitz (1956)

    c) Hansen and Hurwitz (1942) d) Thompson and Hurwitz (1945)

Key to MCQ's

1.	(d)	2.	(c)	3.	(a)	4.	(a)	5.	(b)
6.	(c)	7.	(a)	8.	(b)	9.	(b)	10.	(a)
11.	(b)	12.	(a)	13.	(a)	14.	(a)	15.	(d)
16.	(c)	17	(a)	18.	(c)				

Let us do

1. Explain the meaning of cluster sampling by using the concept of agricultural sciences.
2. In what conditions do we prefer to use cluster sampling over any other sampling technique?
3. What important points should be kept in mind when we perform cluster sampling in any survey?
4. What are the advantages and disadvantages of cluster sampling? Explain in detail with comparison to stratified sampling.
5. What is the process for obtaining the farmer's average income in a district?
6. Explain the difference between cluster sampling and stratified sampling.
7. Explain the process of single-stage cluster sampling with the help of a suitable example.

8. Write down various steps to be followed while using multistage cluster sampling.
9. Explain in brief the concept of area sampling by citing a suitable example.

7

Some Other Sampling Techniques

Multistage Sampling

Multistage sampling is a form of random sampling where samples are selected in a sequence of stages. Where each sample is drawn from the exclusive previously selected samples. This sampling is not done once but at multiple stages. It differs from cluster sampling in the sense that cluster uses all the subjects within a cluster whereas multistage selects a sample from within the selected group. Groups are randomly chosen from a population; subgroups from these groups are randomly chosen and then randomly chosen to be surveyed.

This is an extension of cluster sampling, which is carried out in multiple stages say two, three or four stages. In the first stage, the universe is divided into some clusters from which certain clusters are selected at random at the first stage samples. In the second stage, the selected first stage samples are again subdivided into some clusters from which again certain clusters are selected at random as the second stage samples. In the third stage, the selected second stage samples are again subdivided into some clusters from which certain clusters are again selected at random as the third stage samples. In this way, the process of division and subdivision of clusters and selection of multistage samples are carried out till the sample size is reduced to a reasonable extent.

Note: If we use simple random sampling at every stage to select sub-sample then it is known as multistage random sampling. But if the sub-sample is selected through purposive sampling, then it is known as non-random sampling.

When we use multistage sampling

i. When a researcher suspects that the clusters may not be representative of the population.

ii. When the population is distributed over a large geographical area.

iii. When it is not feasible in terms of cost and time to survey whole clusters or to use simple random sampling.

iv. When it may be expensive to include everyone in the cluster.

v. When it is difficult to divide our population into strata.

Steps in conducting multistage sampling

This method is called multistage sampling because it involves selecting a sample in at least two stages.

i. In the first stage, large group groups or clusters are selected. These clusters are designed to contain more population units that are required for the final sample.

ii. In the second stage, population units are chosen from selected clusters to derive a final sample.

iii. If more than two stages are used, the process of choosing population units within the cluster continues until the final sample is achieved.

Advantages

i. No sampling frame of the population is required.

ii. Most feasible approach for large populations.

iii. The method reduces most of the costs associated with large amounts of data i.e., travelling and listing costs.

iv. Considering the estimates by other methods of probabilistic sampling, multistage sampling reduces variability.

v. It helps to analyze a large population of data that would be very difficult to analyze by using any other method.

vi. Taking the sample several times in a random manner makes this technique worthwhile.

vii. More précised sample observation.

viii. The researcher has the freedom to use n number of stages to reduce the sample size up to the precise, desired extent.

Disadvantages

i. It is less accurate than simple random sampling.

ii. This method is sometimes accounted subjective because of the researcher's freedom of decision-making.

iii. It can lose the external validity of research findings due to subjectivity in sampling.

iv. Several sampling lists are required so this method needs more manpower.

v. If the entire population is not counted this may lead to bias since there is data left out from the unsampled part of the population.

vi. The other probabilistic methods give less error than multistage sampling.

vii. Sampling errors tend to be increased with an increase in the number of stages, so the sampling errors are large in this technique.

Practical Examples

i. If we have to know the impact of PMFBY on farmers in Himachal Pradesh then the multistage random sampling technique will be used to select the ultimate respondents i.e., beneficiaries and non-beneficiaries of PMFBY. The present study includes districts Kangra, Hamirpur, Bilaspur and Una districts of Himachal Pradesh which are selected randomly out of twelve districts in the state. In the first stage, two blocks from each district will be selected randomly. In the second stage from each block, 10 villages will be selected randomly. In the third stage, the list of beneficiaries of the scheme will be obtained from the local bank branch. A total sample of 200 beneficiaries and 200 non-beneficiaries farmers from the size groups i.e., small, medium and large will be selected randomly.

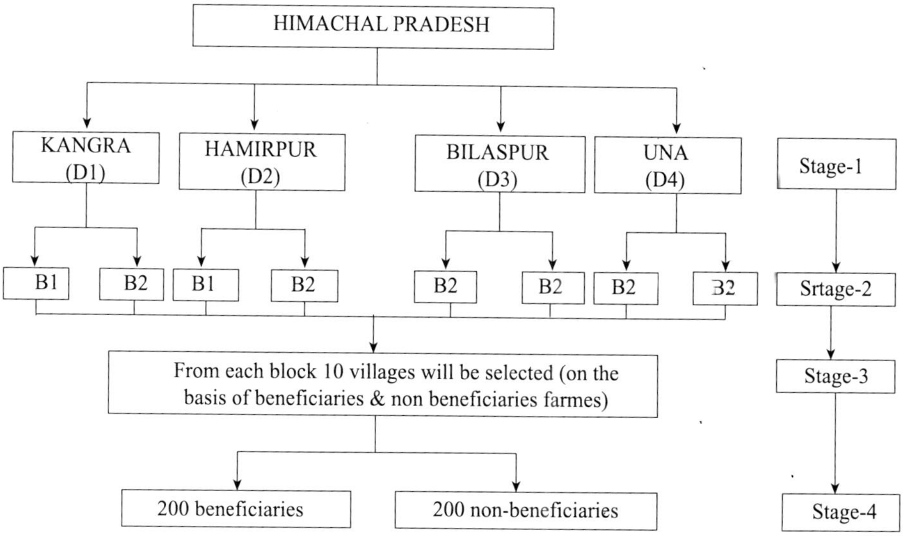

ii. If want to work out the Women's labour participation in Farming Systems of the low and mid hills of Himachal Pradesh. In this case, a multistage random sampling procedure will be used to select the study area and the respondent households. In the first stage three districts, i.e., Kangra and Hamirpur of low hill and Mandi district from the mid-hill zone will be selected randomly among all districts of the state. Thereafter, in the second 50 per cent blocks from each selected district will be selected randomly. In stage three the list of all villages in the selected blocks will be prepared and 60 per cent of the villages will be selected randomly from each block. Ultimately, the fourth stage involves a selection of 30 per cent of the household practising agriculture. The final sample of 250 households representing landless, small, medium and large categories in proportion to the number of total such households in each village will be used for the study.

Multiphase Sampling

There is a thin line of difference between multistage and multiphase sampling. Multiphase sampling ,involves the collection of vital information from the main sample and subsequent collection of more comprehensive information from the small subsamples derived out of the main sample at the same or different points in time. It involves narrowing down the previous sample size based on screening tests at every level. Here, the choices are purposeful and give meaningful results at each level.

Two-phase sampling /Double sampling

In two-phase sampling, sampling is done in two phases, the missing information on an auxiliary variable is gathered in the first phase. Whereas, the second phase involves extensive observations of the study variable chosen from the first phase sample. It is called multiphase sampling when the entire sampling procedure is accomplished in three or more phases.

In the case of two-phase sampling, the size of the second sample on which we observe the study variable will be reduced for the total fixed budget. This is because we had to spend a part of the budget to observe an auxiliary variable on the initial sample which could otherwise be used to observe the study variable on a comparatively larger sample. The technique is therefore beneficial only if the gain in precision is more than the loss in precision due to a reduction in the size of the initial sample.

Merits of multiphase sampling

i. It is a cheaper method.

ii. More purposeful.

iii. It reduces the burden on the part of the answerer.

Example

Suppose there are 100 applicants for the post of Assistant Professor in a university so we have a population of 100 individuals. Therefore, if we apply the multiphase sampling method then, in the first phase a written test was conducted and 40 individuals cleared the exam. Those of passed the first phase are eligible for the second phase which is an interview and after the interview 10 Assistant Professors will be hired by the university. So, in this example sampling units will be candidates for AP's job at every phase.

Difference between Multistage and Multiphase sampling

i. In multiphase sampling, it is necessary to have a complete sampling frame of units. However, the list of the sampling units for the very next stage is required in multistage sampling.

ii. In multistage sampling, sampling units will be different at different stages. On the other hand, multiphase sampling units will be the same at every phase. For example, if the sampling unit is school teachers in a survey, then in two-phase sampling, the samples of teachers will be observed in both phases, whereas multistage sampling involves taking schools as sampling units in stage one and a sample of teachers in the second stage.

Sequential Sampling

It is a non-probabilistic method of sampling technique, initially developed as a tool for product quality control. It is a type of sampling in which the population elements are sampled sequentially, data collection and analysis are done at each stage, and a decision is made as to whether additional population elements should be sampled. The sample size, say 'n' is not fixed in advance, nor is the time frame or date of collection. The process begins first, with the sampling of a single observation. These are then tested to see whether or not the null hypothesis can be rejected. If the null hypothesis is rejected, then another observation or group of observations is sampled and the test will be run again. In this way, the test continues until the researcher is confident in his/her results. A sampling technique in which a certain population is sampled twice is called the double sampling method.

Sequential sampling is an extreme case of multiple sampling that might continue until the lot is exhausted. It calls for verification on a unit-by-unit basis. The final decision is made after each item is inspected for making decisions on rejection or acceptance of the given lot. The sampling and decision-making process was kept on performing until a definite decision is obtained regarding the acceptance or rejection.

Where to use sequential sampling

i. Where sample size is not fixed in advance.

ii. Where the time frame is also not fixed in advance.

iii. Best suitable for comparative studies.

Example

Sequential sampling/sequential modelling is best used to test the quality control in food reserves, water purity and industrial products. It doesn't require sampling at the same point nor does it require large samples at any particular time.

Successive Sampling

Many times, researchers are interested in the time series behaviour of a certain population variable of interest by observing its values over different points in time and also the trends in mean values of that variable in a certain time frame. This information collected on previous occasions can be used to study the change or the total value over occasion for the character, in addition to studying the average value for the most recent occasion.

On recurrent sampling of the population, the chance of modification in the sampling procedure substantially increases. When the same population is sampled repeatedly, the opportunities for a flexible sampling scheme are greatly enhanced. For example, on the nth instance, some characteristics correlate with both $(n - 1)^{th}$ and $(n - 2)^{th}$ occasions, etc. These incidences are quite obvious and are the results of the replacement of the sampling units before the next selection. If the sampling is done on successive occasions according to a specific rule, with partial replacement of sampling units, it is known as rotation sampling.

Objectives

i. To estimate the change with the view to study the effects of the forces acting upon the population. For this, it is better to retain the same sample from occasion to occasion.

ii. To study the overall average or the total. For this, it is better to select a fresh sample for every occasion.

iii. More reliable estimates are expected in the case of the sampling units being retained in subsequent selections compared to other substitutes.

Demerits

i. Repeated surveys of the same sampling units may invite some personal biases and the desired information may not be ascertained accurately.

ii. Repeated surveys may bring about some modification in these sampling units compared to the rest of the population.

Sampling on two occasions

Let N = No. of units in the population

N = No. of units selected on both occasions and remains the same (the value of unit changes over the occasions)

Let us assume that a sample of size 'n' is drawn by SRSWR from the population on the first occasion. Out of this sample of size n, a portion np (= m) of the sample is retained (matched) on the 2nd occasion and supplemented by an additional sample size (unmatched) nq (= u) selected from the remaining (N-n) units, where q is the replacement fraction subjected to p + q = 1.

Example

In milk yield surveys one may be interested in estimating:

i. The average milk yield for the current season.

ii. The change in milk yield for two different seasons.

iii. Total milk production for the year.

Multiple Choice Questions (MCQ's)

1. Suppose you want to investigate the impact of national agricultural policy on Indian farmers, which sampling technique would you follow?

   a) Area sampling b) Multistage sampling

   c) Quota sampling d) Sequential sampling

2. The type of sampling wherein our sample is selected in multiple numbers of stages is known as:

   a) Multistage sampling b) Multiphase sampling

   c) Quota sampling d) Purposive sampling

3. The selection of sub-sample in multistage random sampling can be done via:

   a) Simple random sampling b) Purposive sampling

   c) Both of these d) None of these

4. Under which of the following circumstances multistage random sampling is not quite advantageous over other sampling techniques:

   a) Widely spread geographical area

   b) Survey of each and every unit in a cluster is expensive

   c) When the stratification of the population is difficult

   d) When the stratification of the population is easy

5. Increase in the number of stages in multistage random sampling tends to have a .... in the extent of sampling errors:

   a) Increases b) Decreases

   c) Constant d) Proportional

6. Which of the following are major limitations of multistage random sampling?

   a) Involves more sampling error as compared to other probabilistic sampling techniques

   b) Require more manpower

   c) Involves more subjectivity in decision making

   d) Involves less sampling error as compared to other probabilistic sampling techniques

7. Multistage random sampling involves the division of the universe into groups and subgroups until the sample size is:

   a) Optimized
   b) Reduced to a reasonable extent
   c) Minimized
   d) Maximized

8. If the sample size in multistage sampling is selected via purposive sampling, it is known as:

   a) Random sampling
   b) Non-random sampling
   c) Stratified random sampling
   d) Systematic random sampling

9. The type of sampling involving the collection of vital information from the main sample and subsequent collection of comprehensive information from sub-samples is:

   a) Multistage sampling
   b) Multiphase sampling
   c) Mixed sampling
   d) Purposive sampling

10. In multistage sampling, the unit at each stage is different, whereas in multiphase sampling unit at each stage is:

    a) Same
    b) Different
    c) Unlike
    d) Proportional

11. The sample size in the final stage in multiphase sampling depends upon:

    a) Total budget
    b) Information required
    c) Type of study variable to be examined
    d) Sampling error

12. The main difference between multistage and multiphase sampling is based upon:

    a) Need of sampling frame
    b) Similarity of sampling frame at each stage
    c) Dissimilarity of sampling frame at each stage
    d) Need of sampling units

13. Which type of sampling is preferred mostly by the food testing authority of a country to test the food quality of industries?

    a) Successive sampling
    b) Multiphase sampling
    c) Multistage sampling
    d) Sequential sampling

14. The type of sampling technique wherein the sample size is not fixed in advance is:

    a) Multiphase sampling
    b) Sequential sampling
    c) Successive sampling
    d) Purposive sampling

15. A kind of multiple sampling wherein the entire lot is verified unit by unit until a final decision regarding selection or rejection of a lot is made is known as:

    a) Sequential sampling
    b) Quota sampling
    c) Multiphase sampling
    d) Purposive sampling

16. An environmentalist tends to test the water purity level at a university, he should preferably employ..............sampling.

    a) Quota
    b) Multiphase
    c) Sequential
    d) Successive

17. The time series behaviour of a demographic variable can be tested via:

    a) Multistage sampling
    b) Sequential sampling
    c) Successive sampling
    d) Judgmental Sampling

18. The successive sampling involving partial replacement of units is known as:

    a) Mixed sampling
    b) Rotation sampling
    c) Simple random sampling with replacement
    d) Simple random sampling without replacement

19. Milk yield survey often involves:

    a) Successive sampling
    b) Systematic sampling
    c) Mixed sampling
    d) Sequential sampling

Key to MCQs

1.	(b)	2.	(a)	3.	(c)	4.	(d)	5.	(a)
6.	(a)	7.	(b)	8.	(b)	9.	(b)	10.	(a)
11.	(a)	12.	(b)	13.	(c)	14.	(b)	15.	(a)
16.	(c)	17.	(c)	18.	(b)	19.	(b)		

Let us do

1. Define Multistage. Under what circumstances multistage random sampling is more preferred over simple random sampling?
2. Suggest what should be the primary and secondary or tertiary sampling units under the following circumstances:
   a) Impact of a watershed project in a state.
   b) Socioeconomic impact of MG NERAGA on beneficiaries.
   c) Socioeconomic status of mushroom growers in the Solan district of Himachal Pradesh.
3. What are the major merits and demerits of multistage random sampling?
4. What points should be kept in mind by a researcher while planning to derive a multistage random sample?
5. What are the major points of difference between multistage and multiphase sampling?
6. What are the merits and demerits of double phase sampling?
7. Define sequential sampling. Cite suitable examples. What are its major objectives?
8. Comment on the statement "sampling frame and time frame in sequential sampling is not fixed in advanced."
9. Why and how sequential sampling is used for product quality control?
10. Define successive sampling. Under what situations it is more preferred. Cite suitable examples.
11. Give merits and demerits of successive sampling.
12. Discuss the procedure for sampling on two occasions.

8

Non-Probability Sampling Techniques

Non-Probability sampling methods are also known as non-random sampling methods or deliberate sampling. These types of methods are reliant on the researcher's ability to select members at random. These sampling methods are not fixed or pre-defined selection processes making it difficult for all elements of the population to have equal opportunities to be included in the sample.

Types of non-probability sampling techniques

1. **Judgmental or Purposive sampling**

This technique involves the purposive selection of units of samples based on the judgement of the researcher. The researcher selects the sample on the basis of his a priori knowledge and experience.

Subcategories of purposive sampling

- **Deviant case sampling/Extreme case sampling:** In this, the researcher will select a particular case with a specific characteristic i.e., the focus is on unusual or deviant cases that occur. The presumption is that by examining extreme cases you will be better able to understand typical cases.

Example

If a researcher wants to study the mental disorder called "profound custodial" who has some anti-social behaviour among the students. This case is deviant and not normal so this method is best suitable in this situation.

- **Critical case sampling:** By studying the extreme cases, you can understand the typical characteristic under study. For an instance, a researcher wants to study some of the market failures, with this market failure, 100 companies failed and he will select the one that failed most or the one that was influenced the most. Through this market failure, he can generalize what will happen with less critical cases.

- **Typical case sampling:** This type is just exactly opposite to critical case sampling i.e., if again there was a market failure of 100 failing companies, he would rather choose this average so that he can find and generalize typical failure that has occurred.
- **Homogeneous sampling:** In this case, the researcher will try to get as much variation within his sample according to a relevant key variable.
- **Heterogeneous sampling:** This is the exact opposite of homogenous sampling, here the researcher has defined his key characteristic but selected those cases that differ just very a little from each other. He will actually study some subgroups of the population that may be based on occupational level.

Example

i. Impact assessment of high-density plantations on the socio-economic status of apple orchardists in Himachal Pradesh. Therefore, in this case, every apple orchardist is not the best respondent for the study. Those respondents will be selected who have adopted this technology for this study as they are the adopters and have undergone the benefits/losses of the technology.

ii. If an investigator wants to study the vocational aspirations of visually challenged children in a wide geographical area, then purposive sampling is the best option.

2. Convenience sampling

Sometimes, it is more comfortable for a researcher to observe the members of the sample who are more easily assessable to him/her. This type of selection is most commonly used where the limitation of time and money is there as it saves time and travelling expenditure of the researcher.

Example

i. A researcher may visit his village and get survey the volunteer localities for a socio-economic survey.

ii. Similarly, for a marketing survey, a student may select his nearest market as it is easily assessable to him.

iii. A professional who is a member of a professional organization and wanted to recruit the interviewees through their contact information available to the members of that organization.

iv. A teacher who wanted to examine the perceptions of the teachers about a policy change and decided to utilize a school within the district he/she worked in to recruit participants.

3. Snowball sampling

This sampling is also known as chain sampling as the samples are selected using networks. It is quite interesting to note that, the members are observed from their circle and locality. This technique is commonly embraced in situations when the population is hidden and it is difficult to identify the number of samples.

Example

i. If a researcher is indented to enquire about the problems encountered by the migrants in a district. It is a very cumbersome task to locate each and every migrant to collect data from enough migrants. But the researcher may take the advantage of knowing at least one or two migrants and they will help him to identify the rest of the sample units to make the desired sample size.

ii. If a researcher wants to sample drug addicts or AIDS patients, in these cases it is difficult to identify the individuals so this type of sampling technique holds good.

iii. A researcher is studying the homeless in a city. It is apparently difficult to find the list of all the details of the number of homeless here. However, it is possible to identify one or two homeless individuals who are willing to participate in the study.

Note:

The initial respondents in all the above examples refer to the names of people who might be the sample for the study by the researcher and as such this method is known as the referral sampling method or chain referral method or chain sampling method.

Demerit

Those people who do not have many links are unlikely to be referred to as such may be missed and hence, this method sometimes shows unreliability.

4. Quota sampling

In this type of sampling, specific variables under study serve as the basis for the selection of units of the desired sample. These specific characteristics decide the quota for the study. In simple terms, it is a modified version of stratified random sampling where the units of the sample are selected non-randomly and the universe is partitioned into mutually exclusive strata from where ultimate respondents are to be observed. The basis of quota is usually age, gender, education, race, religion and socio-economic status.

Example

To obtain information on the socio-economic status of 60 farmers above the age of 45, the surveyor has the liberty to select the persons so long as they fall into this age group.

Merits

i. This sampling technique is widely used in opinion or attitude surveys.

ii. Useful if the time and money is given are limited.

Demerit

i. Personal biases are involved.

5. Dimensional Sampling

This is one of the add-on versions of quota sampling. This method takes into account the characteristics of the population such as age, gender, income level, area of living, family size and education etc. while selecting the units to make the sample a better prototype of the population. In this type, after selecting farmers above 45 years of age, the researcher would select on the basis of gender i.e., 50 per cent of them should be males and 50 per cent should be females.

Advantages of Non-Probability sample

i. This sampling is very cost-effective and time-effective.

ii. The method is quite simple, easy and best suited if the given population is not very large and probability sampling is most effective in case of large population size.

iii. These are more useful in pilot studies.

iv. More reliable in the case of social science research, behavioural research and humanities.

Disadvantages of Non-Probability sample

i. These are non-scientific methods.

ii. It involves personal biases.

iii. This method is more prone to sampling errors.

iv. Knowing the efficacy of sampling is very difficult in this method.

v. Calculation on confidence intervals and margins of error is a difficult task.

Differences between probability and non-probability sampling

Basis for comparison	Probability sampling	Non-probability sampling
Meaning	Probability sampling is a sampling technique, in which the subjects of the population get an equal opportunity to be selected as a representative sample.	Non-probability sampling is a method of sampling wherein, it is not known which individual from the population will be selected as a sample.
Basis of selection	Randomly	Arbitrarily
Opportunity of selection	Fixed and known	Not-specified and unknown
Research	Conclusive	Exploratory
Results	Unbiased	Biased
Inferences	Statistical	Analytical
Hypothesis	Tested	Generated
Method	Objective	Subjective
Resources required	More resources like time, cost and efforts	Fewer resources are required

Multiple Choice Questions (MCQ's)

1. The alternate term for non-probability sampling is:

   a) Deliberate sampling b) Random sampling

   c) Mixed sampling d) Systematic sampling

2. The non-probability sampling technique involves...........opportunities for the population elements to be included in the sample:

   a) Unequal b) Equal

   c) Lesser d) Greater

3. Which of the following is not a type of non-random sampling technique?

   a) Quota b) Purposive

   c) Snowball d) Cluster

4. Which one of the following is not a sub-type of purposive sampling?

   a) Critical case sampling b) Typical case sampling

   c) Rare case sampling d) Extreme case sampling

5. Which of the following statements is not true about purposive sampling?

   a) Selection of sample is based on a priori expectations of the researcher

   b) Personal judgment of the researcher

   c) Every unit has an equal probability of being included in the sample

   d) Biasness

6. A Phyto chemist intends to work on anticancerous photochemical in *Taxus Baccata* specifically, he is doing...........case sampling.

   a) Typical b) Critical

   c) Homogeneous d) Heterogeneous

7. The famous writer "Mrs. Amrita Pritam" studied female victims of domestic violence while writing her book Pinjar, this is an example of:

   a) Deviant case sampling b) Typical case sampling

   c) Critical case sampling d) Homogeneous sampling

8. A researcher wants to know the mental ability of class 12th students. For this, he chose his sample from the mediocre students in class instead of brilliant and below-average students. This is a case of........................ sampling.

   a) Typical case b) Critical

   c) Extreme case d) Homogeneous

9. Suppose a pathologist wants to study the disease incidence in the Brinjal crop in his village particularly. This is a case of...........sampling.

   a) Purposive sampling b) Convenience sampling

   c) Dimensional sampling d) Quota sampling

10. Snowball sampling involves sample selection using the............ technique.

    a) Observation b) Networking

    c) Interviewing d) Behavioural survey

11. The snowball technique is most preferable in situations where the variable under study in the population is......... .....

    a) Readily available b) Hidden

    c) Common d) Directly related

12. Hypothesis in probability sampling techniques are tested, but in non-probability sampling techniques hypothesis are...............

    a) Generated b) Tested with low accuracy

    c) Tested with precision d) Not tested.

13. The alternate term for snowball sampling is:

    a) Chain sampling b) Ice ball sampling

    c) Convenience sampling d) Network sampling

14. The modified form of stratified sampling where units of the sample are selected non-randomly from mutually exclusive groups/strata is known as:

    a) Snowball sampling b) Quota sampling

    c) Convenience sampling d) Stratified sampling

15. If a researcher is willing to survey the mental state of patients in a particular district who are below 30 years of age, he should prefer:

    a) Convenience sampling b) Quota sampling

    c) Stratified sampling d) Purposive sampling

16. In a survey on cancer patients below 30 years of age the researcher has acquired additional information on family size and education level also, this type of sampling is known as:

    a) Dimensional sampling b) Deliberate sampling

    c) Quota sampling d) Convenience sampling

17. Social sciences research and behavioural studies mostly involve in.............. sampling.

    a) Probability b) Non-probability

    c) Mixed sampling d) Complete enumeration

Key to MCQs

1.	(a)	2.	(a)	3.	(d)	4.	(c)	5.	(c)
6.	(b)	7.	(a)	8.	(a)	9.	(b)	10.	(b)
11.	(b)	12.	(a)	13.	(a)	14.	(b)	15.	(b)
16.	(a)	17	(b)						

Let us do

1. Define non-probability sampling techniques. Cite various examples from an agricultural research background.
2. Snowball sampling is also termed "chain referral sampling". Comment.
3. Define purposive sampling. Also, discuss various forms of purposive sampling through various examples.
4. Non-probability sampling techniques are more likely to cause subjective biases on the part of research but are necessary evil under certain conditions. Explain why.
5. Differentiate between probability and non-probability sampling in detail.
6. Differentiate between quota sampling and dimensional sampling in detail.
7. Discuss various merits and demerits of non-probability sampling.

9

Sampling with Varying Probabilities

Yet now we have discussed probability sampling techniques in which sampling units are assigned with equal probabilities. But sometimes in practice if want to draw a sample with unequal probabilities i.e., every sampling unit is assigned with varying probabilities and we draw a sample from these units with the help of equal probability methods, then the results are not as much efficient as compared to unequal probability sampling.

Simple random sampling is a procedure wherein the chance of selection of every unit in the required sample is equal. However, the use of simple random sampling is not advisable when the units differ substantially in size as the significance of unit size is ignored.

The basic idea of this type of probability technique is that chance of each unit of a population being selected in the sampling is proportional to its size. The units vary significantly in size and the variable under study is largely related to the size of the unit. Therefore, the probabilities fixed for each unit are proportional to their sizes. In varying probability sampling or probability proportional to size sampling units are selected with probabilities proportional to some auxiliary variable (x). If there are two or more auxiliary variables then we select the one which has a maximum association with the study variable (y).

Example

If we have different agricultural fields and the field sizes are not the same i.e., some fields are smaller in size and some are larger in size. From this, we expect that the smaller field size or area has less production and from a larger field size the production is high. Generally, it happens that the large size plots are less in number and small size plots are more in number. If we apply equal probability sampling in that case, so the large-sized plots may provide lesser representation. If we estimate the overall production from the field then it may not become actual production, because the quantity of larger plots is less and production is high. In that situation, we assigned the maximum weightage to larger-sized plots and vice-versa. When sizes are unequal from where the sampling units are to be selected this sampling method is most commonly used and is also known as probability proportional to size sampling.

Hansen and Hurwitz (1943) were the first who discussed the concept of unequal probability sampling theory. The most commonly used theory is probability proportional to size (PPS).

Probability Proportional to Size (PPS) sampling with replacement

The estimator for population total Y as suggested by Hansen and Hurwitz (1943) is:

$$y'_{HH} = \frac{1}{n}\sum_{i=1}^{n}\frac{y_i}{P_i}$$

Where, Y = auxiliary variable and P = probability of selection

Unbiased Hansen and Hurwitz estimator:

$$E(y'_{HH}) = \frac{1}{n}\sum_{i=1}^{n}E\left(\frac{y_i}{P_i}\right)$$

$$E\left(\frac{y_i}{P_i}\right) = \sum_{i=1}^{N}\frac{Y_i}{P_i}\,P_i = Y$$

Variance of Hansen and Hurwitz estimator:

$$V(y'_{HH}) = \frac{1}{n}\sum_{i=1}^{N}P_i\left(\frac{y_i}{P_i} - Y\right)^2$$

or

$$V(y'_{HH}) = \frac{1}{n}\left(\sum_{i=1}^{N}\frac{Y_i^2}{P_i} - Y^2\right)$$

Example

In our university, there are various departments and in each department the number of faculty is different.

Let us consider six departments and their faculty size.

Department	Assigned Number	Faculty size (Y_i)	Probabilities (P_i)	Assigned Random Number
Management	1	15	15/100	1 - 15
Statistics	2	10	10/100	16 - 25
Economics	3	12	12/100	26 - 37
Horticulture	4	32	32/100	38 - 69
Forestry	5	24	24/100	70 - 93
Computer Sciences	6	7	7/100	94 - 100
Total		100		

Probability Proportional to Size (PPS) sampling without replacement

Let a sample of two units is selected from a population of N units and the probability of selection of the ith unit is $P_i = \frac{Z_i}{Z}$.

Suppose the ith unit is not selected at the first draw but the jth unit is selected ($j \neq i$) then the probability of selecting the jth unit at the first draw is $P_j = \frac{Z_j}{Z}$. The conditional probability of selecting the ith unit at the second draw is $\frac{Z_i}{Z - Z_j} = \frac{P_i}{1 - P_j}$

The probability of inclusion of ith unit at the second draw to be included in the sample is the sum of the product that the jth unit is selected at the first draw and the ith unit is selected at the second draw given the jth unit is selected at the first draw i.e., $\sum_{j \neq i}^{N} P_j \frac{P_i}{1 - P_j}$

The total probability the probability of inclusion of the ith population unit to be in the sample is: $\pi_i = P_i + \sum_{j \neq i}^{N} P_j \frac{P_i}{1 - P_j} = P_i \left[1 + \sum_{i=1}^{N} \frac{P_i}{1 - P_j} - \frac{P_i}{1 - P_j} \right]$

The probability that both i^{th} and j^{th} units are in the sample is denoted by π_{ij} and is defined as the probability of the selection of the ith unit × probability of selection of jth unit given the ith unit is selected at the first draw + probability of selection of jth unit × probability of the ith unit given j^{th} unit is selected at the first draw.

The general theory of unequal probability sampling without replacement was firstly given by Horvitz and Thompson (1952). An unbiased estimator suggested by them for population total Y is

$$Y'_{HT} = \sum_{i=1}^{n} \frac{y_i}{\eth_i} = \sum_{i=1}^{N} a_i \frac{y_i}{\eth_i}$$

Example

ii. Estimation of the total number of unemployed youths in a block (y), for this the auxiliary variable (x) may be the number of households in the block.

iii. Estimation of the total number of irrigation structures in a certain district (y), the auxiliary variable may number of irrigation structures during the previous year (x) or the net irrigated area in the district (x_1).

iv. To select a given sample of orchards has the probability proportional to the number of trees in the orchards. If we are provided with all orchards along with a list of all orchards along with their respective number of trees.

In the case of PPS sampling, the sampling units are selected via simple random sampling with replacement or simple random sampling without replacement. In general, PPSWOR is a more advisable method of sampling over PPSWR. But the procedure of selecting samples under the former is hefty. A lot many procedures are being used to select samples under this method. The main two procedures of selecting samples under the PPS sampling technique are:

i. Cumulative total method

ii. Lahiri's method

1. Cumulative total method

Procedure of cumulative total method

i. Write the cumulative totals for sizes x_i, I = 1, 2, …., N.

ii. One random number r is selected, such that, $1 \leq r \leq x$

iii. Select i^{th} population unit if $T_{i-1} < r \leq T_i$ where $T_{i-1} = x_1 + x_2 + ... + x_{i-1}$ and $T_i = T_{i-1} + x_i$

Drawbacks

The process involves writing down the successive cumulative totals. This is time-consuming and cumbersome if the number of units in the universe is infinite.

Example

Estimate the total number of unemployed persons in 14 villages falling within 20-kilometer radius of the proposed site for a firm. The number of households in these 14 villages are; 400, 350, 760, 432, 860, 1180, 600, 1320, 490, 1040, 310, 520 and 900 respectively. Select the sample of 5 villagers using varying probability without replacement sampling, the size being the number of households.

Solution

Village	Size(x_i)	Cumulative size (T_i)
1	400	400
2	350	750
3	760	1510
4	432	1942
5	860	2802
6	1180	3982

Village	Size(x_i)	Cumulative size (T_i)
7	530	4512
8	600	5112
9	1320	6432
10	490	6922
11	1040	7962
12	310	8272
13	520	8792
14	900	9692

Select r such that, $1 \le r \le 9692$

Let r chosen to be 4479

Here, $3982 \le 4479 \le 4512$

Therefore, 7th unit is selected.

The next four random numbers will be;

$0 \le 191 \le 400$ (1st unit)

$8272 \le 8710 \le 8792$ (13th unit)

$4512 \le 4656 \le 5112$ (8th unit)

$8272 \le 8493 \le 8792$ (13th unit)

2. Lahiri's method

i. Select a random number (let it i) from 1 to N.

ii. Select another number (say j), such that $1 \le j \le m$, m is either equal to the maximum sizes (x_i), i = 1, 2, 3, …, N and the maximum size in the population is more than r.

iii. If $j \le x_i$, i^{th} unit is selected, otherwise, the pair (i, j) of random numbers is rejected, and another pair is chosen by repeating steps i and ii.

Example

Select a sample by varying probability with replacement of size 2 from the population by using Lahiri's method.

Solution

Here N = 14, n = 2 and m = 1320

i. If the value of this m is not available, one could take m equal to 2000, 3000 or more.

ii. We select random number i, $1 \leq i \leq 14$, and then a second random number j, $1 \leq j \leq 1320$, is to be selected

iii. Let be 1st pair (1, 1282) but $1282 > x_1$ (= 400), the first pair of random numbers is rejected.

iv. The next pair (11, 47) results in the selection of 11th village in the sample. The process is repeated until the required sample size is obtained.

Multiple Choice Questions (MCQ's)

1. The other term most commonly used for sampling with varying probabilities is:

   a) Proportional Stratified sampling b) Probability proportional to size

   c) Systematic sampling d) Cluster sampling

2. The technique most commonly used when sampling units vary considerably in size:

   a) Cluster sampling b) Probability proportional to size

   c) Multistage sampling d) Systematic sampling

3. In PPS sampling the elements can be selected via:

   a) Simple random sampling with replacement only

   b) Simple random sampling without replacement only

   c) Both simple random sampling with replacement and simple random sampling without replacement

   d) Purposive sampling

4. The concept of sampling with varying probabilities was first discussed by:

   a) Hansen and Hansen (1942) b) Hansen and Hurwitz (1943)

   c) Lahiri (1943) d) Rao (1940)

5. In sampling with varying probabilities the units are so chosen that their probabilities of selection are proportional to some...............

   a) Study variable b) Auxiliary variable

   c) Random variable d) Non-random variable

6. In the case of PPSWR, the efficiency of the estimates is generally......... than PPSWOR.

   a) More b) Less

   c) Same d) Slightly greater

7. In sampling with varying proportions the selection of samples via PPSWR can be done via:

   a) Cumulative total method only b) Lahiri method only

   c) Desraj method d) Both cumulative total method and Lahiri method

8. A researcher wants to investigate the impact of a government scheme on the beneficiaries of an insurance scheme in different districts of a state but the number of beneficiaries in each district varies so, he should select a sample from the randomly selected districts using .... sampling.

a) PPS b) Stratified sampling

c) Multistage sampling d) Simple random sampling

Key to MCQ's

1.	(b)	2.	(b)	3.	(c)	4.	(b)	5.	(b)
6.	(b)	7.	(d)	8.	(a)				

Let us do

1. Define the term sampling with varying probabilities. What are its advantages over simple random sampling?
2. Discuss in detail the procedure of sample selection by the cumulative total method of selection in PPS with replacement. What are the shortcomings of this method?
3. Write the steps involved in Lahiri's method of sample selection in PPS sampling. Mention the situations when this method is more preferred over the cumulative total method.
4. Give the formulae for population total, population mean and their variances for PPS with replacement.
5. Given the total number of farmers in 10 blocks of Himachal Pradesh and their respective land holding sizes.

Blocks	1	2	3	4	5	6	7	8	9	10
Farmers	90	39	58	69	88	79	60	78	88	90

Select a PPS with replacement sample of 5 blocks, taking the number of farmers as an auxiliary variable via:

i) Cumulative total method

ii) Lahiri method

6. Give some applications of Probability proportional to size sampling in agricultural research.

10

Sampling and Estimation of Proportion

Introduction

In some practical field situations, the results are described by some percentages or proportions. In agricultural field conditions, if any disease prevalence is to find then in the form of percentage or proportion, we can say how much disease rate is present i.e., 10 % or may be any other. So, in those situations, we need to describe our results in proportions. Similarly, in many fields such as medical sciences, social sciences and other allied sciences this type of results in proportions we need to describe.

Estimation

The concept of estimation is derived from the word estimator and these estimators are generally used in the field of statistics that how to draw the inference about the population or target population from the studied sample. Estimation refers to the process by which one makes inferences about a population, based on information obtained from a sample. Statisticians used sample statistics to estimate population parameters i.e., sample means are used to estimate population means and sample proportions are used to estimate population proportions.

Estimator

It is a rule, function and formula that tell how to calculate an estimate.

Estimate

An estimate is the numeric value of the estimator.

Estimation is a process in which we obtained the values of unknown population parameters with the help of sample data.

Some important notations:

For sample data	For population data
n = sample size	N = population size
$\bar{x}$ = sample mean	μ = population mean

S^2 = biased sample variance	σ^2 = population variance
s^2 = unbiased sample variance	
S = sample standard deviation	σ = population standard deviation
P = sample proportion	π = population proportion

Standard error

The standard error (SE) of the sample mean refers to the standard deviation of the distribution of the sample means.

- The standard error gives an estimate of the variability expected if multiple samples are to be drawn from the same population.
- While the standard deviation measures the variability obtained within one sample, the standard error gives an estimate of the variability between many samples.
- Provided the population standard deviation, σ is known, the standard error of the sample mean is denoted as σ_x and is given by is $\sigma_x = \frac{\sigma}{\sqrt{n}}$

Importance of the standard error of the sample mean

- It gives an idea to the statisticians or researchers that how precisely the sample mean estimates the population mean.
- A lower value of the standard error indicates that the estimation of the population mean is more precise.

Point estimation

When an estimate for the unknown population parameter is expressed by a single value it is called a point estimate. A point estimate of a population parameter is a single value of a statistic. i.e., the sample mean $\bar{x}$ is a point estimate of a population mean μ and similarly the sample proportion P is a point estimate of the population proportion π.

Example

Let us assume that a random sample of 10 fruit trees of apple is to be selected from a fruit-bearing apple orchard. The number of fruits from each tree is to be counted as 12, 10, 8, 14, 15, 18, 11, 9, 13, and 10. Calculate the point estimate for (i) population mean (ii) population standard deviation and (iii) standard error of the mean.

Solution

(i) Here, sample size n = 10 and sample mean $\bar{x} = \frac{\sum X}{n} = \frac{120}{10} = 12$

So, the point estimate of the population mean (μ) is 12 and $\bar{x}$ is an estimator.

(ii) Here, the sample standard deviation is $s = \sqrt{\frac{\sum X^2}{n} - \left(\frac{\sum X}{n}\right)^2}$

X	X^2
12	144
10	100
8	64
14	196
15	225
18	324
11	121
9	81
13	169
10	100
$\sum X=120$	$\sum X^2=1524$

$$\therefore\ S = \sqrt{\frac{1524}{10} - \left(\frac{120}{10}\right)^2} = 2.89$$

So, the point estimate of the population standard deviation σ is 2.89 and S is the sample standard deviation.

(iii) The sample standard error of the mean is $\sigma_{\bar{x}} = \frac{\sigma}{\sqrt{n}}$ i.e., $\frac{2.89}{\sqrt{10}} = 0.91$

So, the point estimate of the population standard is 0.91.

Interval estimate

When an estimate for the unknown population parameter is expressed by a range of values within which the population parameter is expected to occur is called an interval estimate. This interval estimate is also defined by two numbers, between which a population parameter is said to lie i.e., $a < \bar{x} < b$ is an interval estimate of the population mean μ. It indicates that the population mean is greater than a but less than b.

Confidence intervals

We cannot study or measure the exact population characteristics there may be a difference between the population parameter and the estimator or estimation. But we tried to minimize or precise as much possible as we can. So, for this, we make an interval in which we can say that the population parameter lies between 5 to 10 which makes interval and we attach a level of confidence with that if it is 95% i.e., we are 95% confident that the parameter lies between 5 to 10 it means in repeated sampling if we estimate the population parameter and which estimator is to come it is 95% of the time falls in the interval of 5 to 10.

Let us consider if we take 100 samples and out of these 100 samples, we obtain the estimates with a 99% level of confidence. Then, the results of 99 samples would fall in that interval which we construct with these 100 samples and their estimate this means that it is a confidence interval.

Confidence interval for population mean and population proportion

For these two things very important first is sample estimate ($\bar{y}$) and p in case of proportion and second is the standard error. Generally, two cases are discussed in the case of mean (i) if population variance is given i.e., known variance (ii) if population variance is not given i.e., unknown variance. When population variance is given then we use standard normal distribution and when the population variance is not given then we use student t-distribution.

The interval estimation of the population mean for known variance is:

$$\bar{y} \pm z_{1-a/2}\ \text{SE}(\bar{y})$$

Lower limit

$$\bar{y} - z_{1-a/2}\ \text{SE}(\bar{y})$$

Upper limit

$$\bar{y} + z_{1-a/2}\ \text{SE}(\bar{y})$$

The interval estimation of the population mean for unknown variance is:

$$\bar{y} + t_{1-a/2,y}\ \text{SE}(\bar{y})$$

Example

The apple juice company produces apple juice bottles with a standard deviation of the amount poured into bottles by an automatic filling machine is 1.5 ml. A random sample is taken of the amount of filling in bottles were 115, 117, 112, 116, 114, 119, 116, 114, 115, 113. Suppose that population of filling amount follows a normal distribution. Determine the 95% confidence interval for the mean amount in all cans filled by the machine.

Solution

Here, we have given $\sigma = 1.5$ (population standard deviation)

Now, according to the formula we need $\bar{y}$ (sample mean) i.e.,

$$\bar{y} = \sum_{i=1}^{n} \frac{y_i}{n} = \frac{1151}{10} = 115.10$$

Therefore, $SE(\bar{y}) = \frac{\sigma}{\sqrt{n}} = \frac{1.5}{\sqrt{10}} = 0.474$

Now, the 95% confidence limits will be $\bar{y} \pm 1.96\ SE(\bar{y})$ here, at 95% the z-value is 1.96

Lower limit = 115.10 - 1.96 (0.474) = 114.17

Upper limit = 115.10 + 1.96 (0.474) = 116.30

Confidence interval for the population proportion

The interval estimation of population proportion is:

$p \pm z_{1-\alpha/2}\ SE(p)$

Example

A random sample of 120 apple plants is taken from a population of 650 plants in the apple growing belt. The total number of infested plants in the sample was 34. Construct a 95% confidence interval.

Solution

Here, the proportion is $p = \frac{34}{120} = 0.28$

Therefore, $q = 1 - p = 0.72$

Now, $SE(p) = \sqrt{\frac{pq}{n}} = \sqrt{\frac{0.201}{120}} = 0.0037$

The 95% confidence limits will be:

$p \pm 1.96\ SE(p)$

Lower limit = 0.28 - 1.96 (0.0037) = 0.272

Upper limit = 0.28 + 1.96 (0.0037) = 0.287

Statisticians used confidence intervals to express the precision and uncertainty associated with a particular sampling method. A confidence interval consists of three parts.

- A confidence level
- A statistic
- A margin of error

The confidence level describes the uncertainty of the sampling method.

The statistic and the margin of error define an interval estimate that describes the precision of the sampling method.

The interval estimate of a confidence interval is defined by the sample statistic margin of error.

Let us consider an example if we want to compute an interval estimate of a population parameter. Then, we have to describe this interval estimate as a 95% confidence interval. This means that if we used the same sampling method to select different samples and compute different interval estimates, the true population parameter would fall within a range defined by the sample statistic ± margin of error 95% of the time.

Confidence intervals are preferred to point estimates because confidence intervals indicate the precision of the estimate and uncertainty of the estimate.

Estimation of proportion

Generally, when we study our characteristics also known as a variable of interest if our interest is to study the fruit-infested rate in an apple orchard. If there is an infestation in fruit or suffering from some disease, so the presence of that disease is assigned by a value of 1 and if any fruit is not infested or not get suffered by disease then we have to assign a value of 0 means that disease is absent.

Let us have a population of size N i.e., $Y_1, Y_2, \ldots, Y_i, \ldots, Y_N$

If the presence of disease is there in any i^{th} fruit the assigned value is 1 otherwise 0. The sum of Y_i's is taken for those units which possess a certain attribute 1.

Note:

$\sum Y_i = \sum Y_i^2$ i.e., the sum of Y_i's is equal to the sum of Y_i^2

Example

There are 6 fruits to be observed i.e., $Y_1\ Y_2\ Y_3 Y_4\ Y_5\ Y_6$ out of these six fruits first, third, fifth and sixth is having disease it means out of six fruits four are infested. In such types of cases when we want to draw a sample then we have to calculate its proportion to have a representative sample.

Description of proportional mean

Mean for population proportion is:

$$\bar{Y} = \sum_{i=1}^{N} \frac{Y_i}{N} = \frac{A}{N} = P$$

Here, $\sum Y_i$ is the number which we get in any case and that is equal to A and similarly in case of the sample proportion.

Mean for a sample proportion is:

$$\bar{y} = \sum_{i=1}^{n} \frac{y_i}{n} = \frac{a}{n} = p$$

Since Y_i takes the values as 1 and 0

$\sum Y_i^2 = A = NP$ same as in case of sample $\sum y_i^2 = a = np$

Description of proportional variance

Variance for population proportion is:

$$\sum_{i=1}^{N} (Y_i - \bar{Y})^2 = \sum_{i=1}^{N} Y_i^2 - N\bar{Y}$$

$$\therefore\ S^2 = \frac{\sum_{i=1}^{N} (Y_i - \bar{Y})^2}{(N-1)} = \frac{N}{N-1} P(1-P)$$

$$\Rightarrow S^2 = \frac{NPQ}{N-1}$$

Similarly, $s^2 = \dfrac{npq}{n-1}$ is used for sample proportional variance

Multiple Choice Questions (MCQ's)

1. Inferential statistics is the:
   a) The process of using a population parameter to estimate the values for the sample statistic
   b) The process of using sample statistics to estimate the population parameter
   c) A process which allows the researcher to determine the exact values for population parameters
   d) A process that eliminates the problem of sampling error
2. A point estimator is defined as:
   a) The average of the sample values
   b) The average of the population values
   c) A single value that is the best estimate of an unknown population parameter
   d) A single value that is the best estimate of an unknown sample statistic
3. An interval estimate is an estimate of the .... of possible values for a population parameter.
   a) Range b) SD
   c) MD d) QD
4. When sample data are used for estimating a population mean .... error will not be present since the observed sample statistic will not differ from the actual value of the population parameter.
   a) Standard b) Sampling
   c) Non-sampling d) Type-I
5. A confidence interval is defined as:
   a) A point estimate plus or minus a specific level of confidence
   b) An interval used to infer something about an unknown sample statistic value
   c) An interval that has a 95 % probability of containing the population parameter
   d) A lower and upper confidence limit is associated with a specific level of confidence

6. The term $1-\alpha$ refers to the:

   a) Level of confidence minus one b) Level of confidence plus one

   c) Level of confidence d) Level of significance

7. A 95 % confidence interval for the population mean is calculated to be 75.29 to 81.45. if the confidence level is reduced to 90 %, the confidence interval will:

   a) Become narrower b) Remain the same

   c) Become wider d) Double in size

8. A 95 % confidence interval for the population mean is calculated to be 75.29 to 81.45. if the confidence level is reduced to 99 %, the confidence interval will:

   a) Become narrower b) Remain the same

   c) Become wider d) Double in size

9. In the formula for the confidence interval, $z_{\alpha/2}$ is part of the formula. What does the sub script $\alpha/2$ refer to?

   a) The area in the lower tail or upper tail of the sampling distribution of the sample mean

   b) The probability that the confidence interval will contain the population mean

   c) The probability that the confidence interval will not contain the population mean

   d) The area in the lower tail or upper tail of the sampling distribution of the population mean

10. Larger the level of confidence used in constructing a confidence interval estimate of the population mean by:

    a) Narrower the confidence interval

    b) Wider the confidence interval

    c) Remain same the confidence interval

    d) Smaller the value of $z_{\alpha/2}$

11. The boundaries of a confidence interval are called:

    a) Confidence level b) Degree of confidence

    c) Confidence limits d) Significance level

12. What value of z would you use to calculate the 80 % confidence interval for a population mean, given that you know the population standard deviation, the sample size and the sample mean of your sample?

    a) 1.96 b) 2.58

    c) 0.80 d) 1.28

13. The width of the interval increases when the confidence level is..................

    a) Increased b) Decreased

    c) Constant d) Half

14. The width of the interval decreases when the significance level is ....

    a) Increased b) Decreased

    c) Constant d) Half

15. The width of the interval increases when the sample size is..............

    a) Increased b) Decreased

    c) Constant d) Half

16. After constructing a confidence interval estimate for a population mean, you believe that the interval is useless because it is too wide. In order to correct this problem, you need to:

    a) Increase the population standard deviation

    b) Increase the level of confidence

    c) Decrease the sample size

    d) Increase the sample size

17. The problem with relying on a point estimate of a population parameter is that the point estimate:

    a) Does not tell us how close or far the point estimate might be from the parameter b) Has no variance

    c) May not be consistent d) May be efficient

Key to MCQ's

1.	(b)	2.	(c)	3.	(a)	4.	(b)	5.	(d)
6.	(c)	7.	(a)	8.	(c)	9.	(a)	10.	(b)
11.	(c)	12.	(d)	13.	(a)	14.	(a)	15.	(b)
16.	(d)	17	(a)						

Let us do

1. Explain the concept of estimation of proportion by citing a suitable example of agricultural field experiments.
2. Write the short notes on the following terms:
   a) Estimation
   b) Estimator
   c) Estimate
   d) Standard error
   e) Point estimation
   f) Interval estimate
   g) Confidence intervals
3. What is the importance of standard error of mean 'SE(m)' in any research? How it is calculated, justify this statement with an example.
4. A random sample of 15 persons is to be selected from a university. The annual income of each person (in Lakhs) is to be counted as 10, 12, 8, 15, 13, 17, 14, 9, 7, 13, 8, 13, 12, 17 and 10. Calculate the point estimate for (i) population mean (ii) population standard deviation and (iii) standard error of the mean.
5. A random sample of 150 apple plants is taken from a population of 700 plants in the apple growing belt. The total number of infested plants in the sample was 50. Construct a 95% confidence interval.

11

Ratio and Regression Estimators

In our previous chapters, we have discussed various sampling designs or techniques such as simple random sampling, stratified sampling, systematic sampling and cluster sampling recently we have discussed unequal probability sampling i.e., probability proportional to size sampling. In all these sampling designs the variable of interest (generally there is only one variable of interest) which we have considered and perform our estimation based on that single variable. There are some other variables which are related to our variable of interest and if we utilize those variables, we may gain some efficiency in our estimator i.e., if we take some correlated variables with our variable of interest or study variable by this, we can enhance our estimation. The variable which is correlated with the study variable is usually known auxiliary variable or supporting variable or supplementary variable or benchmark variable. This auxiliary variable is utilized in two stages either in the design stage or estimation stage.

Graunt (1662) was the first who used auxiliary information to estimate the population size of England. After this Laplace used auxiliary information for the estimation of the population in France.

Example for explanation

In Himachal Pradesh there are 12 districts and all the births are registered in these districts. We are interested to estimate the total population of the state. The registered births have a relation to the population size.

The relationship which we have discussed is further extended to a linear relationship i.e., the correlation between study variable (y) and auxiliary variable (x) when plotted on a graph provides a linear line also known as a direct relationship.

Let $y \propto x$ which means both the variables have a positive correlation. When the line passes through the origin such a type of trend occurs then we use the ratio type of estimation. When the line doesn't pass through origin with available intercept in such type of situations, we use a regression type of estimator.

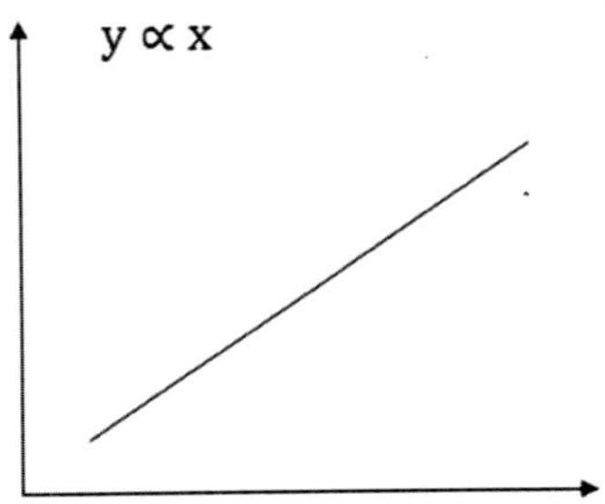

Ratio Method

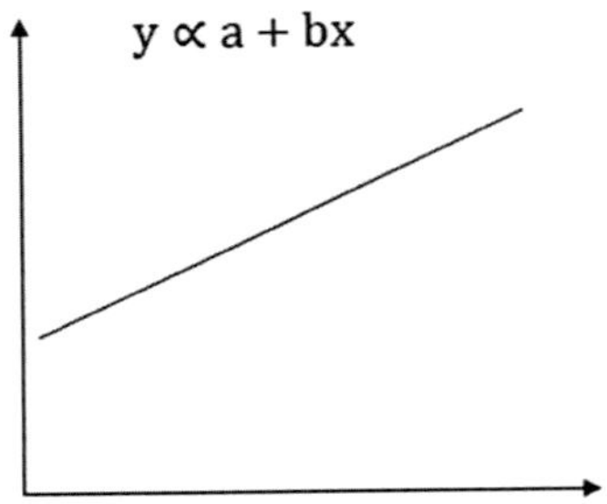

Regression Method

Ratio Estimator

This method of estimation is used when we have a study variable and also, we have an auxiliary variable to be used in support of the study variable. In this method, the auxiliary variable is considered as a variable that is already available to us moreover no efforts and resources are required to get this variable.

We know that in the case of ratio method of estimation, we have

$y \propto x$

therefore, we can write it as $y = R\,x$ by introducing a constant R

in case of means $\bar{y} = R\,\bar{x}$

$\therefore R = \dfrac{\bar{y}}{\bar{x}}$

Ratio estimator is $\hat{R} = \dfrac{y}{x} = \dfrac{\bar{y}}{\bar{x}}$

Ratio estimator for the population total is $y_r = \dfrac{y}{x} X$

Ratio estimator for the population mean is $\bar{y}_r = \dfrac{\bar{y}}{\bar{x}} \bar{X}$

Ratio estimator is a biased estimator but as the sample size increases the biasness of this estimator is removed. So, we can say that for the large sample size this estimator is approximately unbiased estimator it means for the large sample size the expectation of ratio estimator is approximately equal to the population ratio i.e., $E(\hat{R}) \approx R$

Regression Estimator

When the relation between the variable under study and the auxiliary variable is linear in nature and doesn't pass through the origin in such a situation regression method of estimation is used. Ratio and Regression estimates are identical when the regression line passes through the origin. A ratio estimate is a special case of regression estimate.

The regression estimator for the population mean is:

$$\bar{y}_{reg} = \bar{y} + \hat{\beta}_{yx} (\bar{X} - \bar{x})$$

Example

Let us consider a hypothetical example of forestry tree species of Devdar with the following data.

Units	1	2	3	4	5	6
Y = volume (in m³)	3.31	4.21	4.87	5.69	6.27	6.94
X = diameter (in cm)	56	61	66	71	75	80

We are interested to take all possible samples of size 2 and also calculate the expected value for the ratio estimator.

Solution

We draw all possible samples with SRSWOR and we know that ratio estimator is $\bar{y}_r = \frac{\bar{y}}{\bar{x}}\bar{X}$ also, we apply this formula when we know the population mean of auxiliary variable i.e., $\bar{X} = 68.17$

Now, we calculate all possible samples without replacement

Units		Y_i's	Y_i's	$\bar{y}$	X_i's	X_i's	$\bar{x}$	$\bar{y}/\bar{x}$	$\bar{y}_r$
1	2	3.31	4.21	3.76	56	61	58.50	0.064	4.38
1	3	3.31	4.87	4.09	56	66	61.00	0.067	4.57
1	4	3.31	5.69	4.50	56	71	63.50	0.071	4.83
1	5	3.31	6.27	4.79	56	75	65.50	0.073	4.99
1	6	3.31	6.94	5.13	56	80	68.00	0.075	5.14
2	3	4.21	4.87	4.54	61	66	63.50	0.071	4.87
2	4	4.21	5.69	4.95	61	71	66.00	0.075	5.11
2	5	4.21	6.27	5.24	61	75	68.00	0.077	5.25
2	6	4.21	6.94	5.58	61	80	70.50	0.079	5.39
3	4	4.87	5.69	5.28	66	71	68.50	0.077	5.25
3	5	4.87	6.27	5.57	66	75	70.50	0.079	5.39
3	6	4.87	6.94	5.91	66	80	73.00	0.081	5.51
4	5	5.69	6.27	5.98	71	75	73.00	0.082	5.58
4	6	5.69	6.94	6.32	71	80	75.50	0.084	5.70
5	6	6.27	6.94	6.61	75	80	77.50	0.085	5.81

$$\therefore E(\bar{y}_r) = \sum \bar{y}_r P(s)$$

$$= 4.38 \times \frac{1}{15} + 4.57 \times \frac{1}{15} + \ldots + 5.81 \times \frac{1}{15}$$

$$= 5.19$$

$$\therefore MSE(\bar{y}_r) = \sum_{i=1}^{15} (\bar{y}_{ri} - \bar{Y})^2 P(s)$$

$$= (4.38 - 5.22)^2 \times \frac{1}{15} + (4.57 - 5.22)^2 \times \frac{1}{15} + \ldots + (5.81 - 5.22)^2 \times \frac{1}{15}$$

$$=0.16$$

Product Estimator

This type of estimator is used when there is a strong and linear relationship between the study variable and auxiliary variable, but this relation is negative or inverse in nature.

Product estimator for the population mean is $\bar{y}_p = \frac{\bar{y}\bar{x}}{\bar{X}}$

Example

A similar example is considered with an inverse relationship and bit changes in values.

Let us consider a hypothetical example of forestry tree species of Devdar with the following data.

Units	**1**	**2**	**3**	**4**	**5**	**6**
Y = volume (in m^3)	3.31	4.21	4.87	5.69	6.27	6.94
X = diameter (in cm)	80	75	71	66	61	56

We are interested to take all possible samples of size 2 and also calculate the expected value for ratio estimator.

Solution

We draw all possible samples with SRSWOR and we know that ratio estimator is $\bar{y}_p = \frac{\bar{y}\bar{x}}{\bar{X}}$ also, we apply this formula when we know the population mean of the auxiliary variable i.e., $\bar{X}$= 68.17

Now, we calculate all possible samples without replacement

Units		Y_i's	Y_i's	$\overline{y}$	X_i's	X_i's	$\overline{x}$	$\overline{y} \times \overline{x}$	$\overline{y}_p$
1	2	3.31	4.21	3.76	80	75	77.50	291.40	4.27
1	3	3.31	4.87	4.09	80	71	75.50	308.80	4.53
1	4	3.31	5.69	4.50	80	66	73.00	328.50	4.82
1	5	3.31	6.27	4.79	80	61	70.50	337.70	4.95
1	6	3.31	6.94	5.13	80	56	68.00	348.50	5.11
2	3	4.21	4.87	4.54	75	71	73.00	331.42	4.86
2	4	4.21	5.69	4.95	75	66	70.50	348.98	5.12
2	5	4.21	6.27	5.24	75	61	68.00	356.32	5.23
2	6	4.21	6.94	5.58	75	56	65.50	355.16	5.36
3	4	4.87	5.69	5.28	71	66	68.50	361.68	5.31
3	5	4.87	6.27	5.57	71	61	66.00	367.62	5.39
3	6	4.87	6.94	5.91	71	56	63.50	374.97	5.50
4	5	5.69	6.27	5.98	66	61	63.50	379.73	5.57
4	6	5.69	6.94	6.32	66	56	61.00	385.22	5.65
5	6	6.27	6.94	6.61	61	56	58.50	386.39	5.67

$$\therefore E\left(\overline{y}_p\right) = \sum \overline{y}_p P(s)$$

$$= 4.27 \times \frac{1}{15} + 4.53 \times \frac{1}{15} + \ldots + 5.67 \times \frac{1}{15}$$

$$= 5.16$$

$$\therefore MSE\left(\overline{y}_p\right) = \sum_{i=1}^{15} \left(\overline{y}_{pi} - \overline{Y}\right)^2 P(s)$$

$$= (4.27 - 5.22)^2 \times \frac{1}{15} + (4.53 - 5.22)^2 \times \frac{1}{15} + \ldots + (5.67 - 5.22)^2 \times \frac{1}{15}$$

$$= 0.16$$

Multiple Choice Questions (MCQ's)

1. Regression method of estimation is used to get:
   a) Small variance b) Large variance
   c) Unbiased variance d) Biased but consistent variance

2. A questionnaire consists of a set of queries about a:
   a) Specific issue b) Sample
   c) Population d) Sample design

3. Regression method of estimation was introduced by:
   a) R A Fisher b) Karl Pearson
   c) Watson d) Cochran

4. In the regression method of estimation β is.
   a) Independent variable b) Dependent variable
   c) Regression coefficient d) Slope of the line

5. Regression estimation can be made in...............
   a) One way b) Two way
   c) Three way d) None of the above

6. The bias is negligible in...........
   a) Large samples b) Small samples
   c) Ratio d) Sample mean

7. In the regression method of estimation, the sample mean is an unbiased estimate of the population...................
   a) Total b) Mean
   c) Ratio d) Variance

8. In regression method of estimation sample variance is smaller than variance of:
   a) Ratio method of estimation b) Stratified random sample
   c) Simple random sample d) All the above

9. Ratio method of estimation is not applicable when regression line is passing.

   a) Not through origin b) Through origin
   c) Through X-axis d) Through Y-axis

10. .... was the first who used auxiliary information to estimate the population size of England.

   a) Laplace (1632) b) Graunt (1662)
   c) Miller (1732) d) Watson (1822)

11. When study variable (y) and auxiliary variable (x) are plotted on a graph provides a linear line that is also known as:

   a) Path relationship b) Indirect relationship
   c) Direct relationship d) Multiple relationship

12. When the line passes through the origin then which type of estimation occurs?

   a) Ratio b) Regression
   c) Product d) Factor

13. When the line doesn't pass through the origin then which type of estimation occurs?

   a) Ratio b) Regression
   c) Product d) Factor

14. When $y \propto x$ which type of estimation method is used:

   a) Ratio b) Regression
   c) Product d) Factor

15. Ratio estimator is a .... type of estimator.

   a) Biased b) Unbiased
   c) Efficient d) Sufficient

16. To remove the biasness of the ratio type of estimator the sample size is to be ....

   a) Increased b) Decreased
   c) Constant d) Precise

17. If there is a strong and inverse linear relationship between y and x ....method of estimation is used.

a) Ratio b) Regression

c) Product d) Factor

Key to MCQ's

1.	(a)	2.	(a)	3.	(c)	4.	(d)	5.	(b)
6.	(a)	7.	(b)	8.	(c)	9.	(a)	10.	(b)
11.	(c)	12.	(a)	13.	(b)	14.	(a)	15.	(a)
16.	(a)	17	(c)						

Let us do

1. In what situations we use the auxiliary information to estimate the population size?
2. Explain the concept of ratio method of estimation with the help of diagrammatical representation. Also, explain a suitable example.
3. Explain the concept of regression method of estimation with the help of diagrammatical representation. Also, explain a suitable example.
4. Explain the concept of product method of estimation with the help of diagrammatical representation. Also, explain a suitable example.
5. Differentiate the key points between the followings:
   a) Ratio vs Regression method of estimation
   b) Ratio vs Product method of estimation
   c) Regression vs product method of estimation
6. Let us consider a hypothetical example of forestry tree species of Chir pine with the following data.

Units	**1**	**2**	**3**	**4**	**5**	**6**	**7**
Y = volume (in m^3)	2.43	3.12	3.96	4.22	4.78	5.17	5.65
X = diameter (in cm)	94	87	81	79	75	69	67

We are interested to take all possible samples of size 3 and also calculate the expected value for ratio estimator.

12

Non-Response and Unconventional Sampling

Non-Response

Non response occurs when respondents are not available or do not occur in the survey and refuse to respond by any means. This non response is mainly of two types either item non response or unit non response.

If we are conducting an online survey and mailing a questionnaire to various persons. But sometimes we get a reply for some questions and sometimes not i.e., in case of questions in which income is concerned. In some cases, the mail is ignored due to lack of time. So, if the whole questionnaire is ignored by any means, then it is known as non-response and if some of the questions are ignored due to their fussy nature such type of response is known as a unit non response. Such types of non-response problems are very common in surveys. Whenever we carry out a survey non response always exists because it is almost impossible to get every response from each individual.

The individual chosen for the sample is not ready to participate in the survey is known as non-response. This type of selection bias is unit/item non response.

The problem of non-response can be determined with the help of the following methods:

• Sub sampling of non-respondents

This method was introduced by Hansen and Hurwitz (1946) according this; if we conduct a survey once completely and we have collected some non-responses then we have to apart these non-responses from response groups. Take a sample from non-respondent group and study their estimates.

• Randomized response technique

This technique is used when we study or estimate the sensitive characteristics of a population. By sensitive characteristics, we understand that those variables to be studied or those questions to be asked from the respondent by which respondent come across to hesitate for such answer. For example, we want to know someone's income or whether someone is a smoker or not. Information regarding tax returns etc. such types of questions is sensitive. Whenever we are studying such characters and respondents get hesitant, not confident and chances to have the wrong reply if the response comes in such situations, we use randomized response technique.

This technique was first introduced by S L Warner (1965) at that time this technique was used to study qualitative characteristics such as if we are interested to study the proportion of smokers in our university. For such characters this technique was used, later on, quantitative characters were also studied by this technique. So, this randomized response technique was used to study both the qualitative and quantitative characteristics and is known as qualitative and quantitative response models.

Qualitative Response Model

This model is used to estimate the proportion of some behaviour or occurrence in a population i.e., to estimate the "proportion of people who smoke a cigarette".

Warner introduced the following model:

$$Z = Y_p + (1-p)\,(1-Y)$$

Where,

Z = Proportion of YES answers given by respondents

Y = True proportion of those interviewed bearing the sensitive character

p = Probability of answering the sensitive questions

This model can be transformed as:

$$Y = \frac{Z + p - 1}{2p - 1}$$

In such a response model we have two statements either true or not true.

e.g., Statement 1: I truly fill the tax returns, Statement 2: I do not truly fill the tax returns.

Statement 1: I smoke cigarettes, Statement 2: I never smoke a cigarette.

Example

In a hypothetical survey regarding the tax returns, in such cases, we have to give two statements to respondents i.e., Statement 1 which is known as a sensitive statement and Statement 2 which is known as an alternative statement. These two statements provide confidence to the respondents to answer the respective question. For this a random number from 1 to 6 with the help of dice or any chits or by any other means, they have to choose a number if 6 comes, they chose statement 1with a response YES or NO and any other number 1 to 5 comes, chose statement 2 either YES or NO.

Statement 1: I have falsified my tax returns.

Statement 2: I have never falsified my tax returns.

Due to this respondent get confident to answer the question because he thinks the interviewer doesn't know for which statement, he chooses YES or NO.

Let us suppose that we have surveyed 100 respondents and the probability comes out for the response of sensitive character is 1/6. Assume that out of 100 respondents 80 say YES for any statement. But the interviewer doesn't know for which statement they say YES.

Therefore, Warner's Model is $Z = Y_p + (1 - p)(1 - Y)$

Now, $\frac{80}{100} = Y\left(\frac{1}{6}\right) + \left(1 - \frac{1}{6}\right)(1 - Y)$

$$\therefore Y = \frac{\frac{4}{5} + \frac{1}{6} - 1}{2\left(\frac{1}{6}\right) - 1} = 0.05$$

It means 5% of respondents falsified their tax returns.

Quantitative Response Model

This model is used for sensitive quantitative characters such as to estimate the monthly income of the head of a family. To estimate the total value of purchased orders in a year for a company. To estimate the number of times students, bunk the class during a semester.

In all our previous chapters we have discussed various conventional sampling designs. In this chapter, we have discussed such types of designs in which the population is hard to detect i.e., data or observations taken from the population are inquisitive. In such situations, we used unconventional sampling designs.

Capture Recapture Sampling

This method of sampling is used when we have to estimate the population size. In this method, we select a sample and mark the selected units with the help of any unique identity or by any tag number. This marking is important because when we select a sample again so that we can easily identify whether it is selected previously or not. After this release back the selected units to the population.

Select a second sample independently from the same population and noted how many samples or units are selected again which are selected in the first sample which was marked and estimate the proportion of the marked units from the second sample. The capture-recapture technique is generally used in ecological surveys for the study of animals and birds.

Notations

T = Number of units in the population

K = Number of units marked on the first sampling instance

n = Number of units captured on the second sampling instance

k = Number of recaptured units that were marked

Since the proportion of the marked units in the recaptured sample is likely to be about the same as the first sample in the whole population.

$$\therefore \qquad \frac{k}{n} = \frac{K}{T}$$

So, the estimated population size is $T = \frac{nK}{k}$

Example

We are interested to estimate the bird population near Pong Dam District Kangra, Himachal Pradesh. An initial sample of 80 birds was selected and marked with the tag number, after this process, all birds were released back to the population to intermingle. A sample of size 150 is selected on the second visit and 40 of them were selected on the first visit.

Here, K = 80, n = 150 and k = 40 so we can easily estimate the population size T.

$$\therefore T = \frac{nK}{k} = \frac{150 \times 80}{40} = 300$$

Hence, in that particular area near Pong Dam, there are approximately 300 birds.

Line and Point Transects

In this method usually, the estimation is used for population density.

(In this book we are not discussing this method in detail)

Multiple Choice Questions (MCQ's)

1. Item non-response is a problem that frequently occurs in............data.

   a) Time series b) Survey

   c) Cross-sectional d) Geographical

2. Non-response occurs when respondents are............to respond by any means.

   a) Refuse b) Agree

   c) Happy d) Decide

3. Which of the following sampling method is not based on probability?

   a) Systematic sampling b) Cluster sampling

   c) Simple random sampling d) Quota sampling

4. The method of sub-sampling of non-respondents is given by Hansen and Hurwitz in the year.

   a) 1956 b) 1946

   c) 1948 d) 1952

5. Randomized response technique is used to study the .... characteristics.

   a) Robust b) Insensitive

   c) Sensitive d) Homogeneous

6. Randomized response technique was first introduced by

   a) Hurwitz and Thompson (1942) b) Hansen and Hurwitz (1948)

   c) S L Warner (1965) d) Lahiri (1968)

7. Qualitative response model given by Warner is.

   a) $Z = Y_p + (1 + p)(1 - Y)$ b) $Z = Y_p + (1 - p)(1 - Y)$

   c) $Z = Y_p + (1 - p)(1+Y)$ d) $Z = Y_p + (1 + p)(1 + Y)$

8. Which of these is a correct assumption you need to make about a population between samplings?

   a) It is a closed population b) It won't rain between samplings

   c) There are lots of deaths d) There are lots of births

9. If 50 crabs were captured and marked and then released and then a few days later 25 were caught and of those 5 were marked. What would the estimate of the population be?

   a) 500 b) 250

   c) 75 d) 200

10. Conservationists want to find out a population of a species of the mountain goat. On one day they capture 12 goats tag them and release them. A week later they capture 8 goats, 5 of which are tagged. What would be the estimate of the population?

    a) 19 b) 40

    c) 72 d) 30

11. I have a box with 300 balls in it. I take 100 balls and mark them, put them back in and mix them up. If I took a sample of 60 balls how many would I expect to be marked?

    a) 150 b) 75

    c) 50 d) 20

12. Using a valid, verified index of population size we can:

    a) Look at the relative increase or decrease in population size between years

    b) Evaluate the magnitude of difference in size of two populations.

    c) Describe the density of a given population.

    d) Compare the density of two populations

13. When sampling from a population, we would like to choose a random sample. Why is a random sample preferable to other sampling schemes?

    a) It gives a more precise estimate of population parameters

    b) It produces unbiased estimates for the population

    c) It allows the biologist to choose the sample units he feels represent the population more accurately

    d) It is one of the easier sampling schemes to implement

14. A mark-recapture estimate of population size will be affected by animals losing their marks as follows.

    a) Too low

    b) Too high

    c) Wrong but sometimes too high and other times too low

    d) Unaffected

15. The process of inferring from what happened in a few cases to what will happen in all cases is termed as.

    a) Induction b) Deduction

    c) Reproduction d) Introduction

16. How likely is it that a density estimate will be within its 95% confidence interval?

    a) 5 % b) 50 %

    c) 95 % d) 100 %

17. If some animals become trap shy, the mark-recapture estimate will be.

    a) Too high b) Too low

    c) Highly variable d) Unaffected

18. An accurate estimate is.

    a) Unbiased b) Biased but precise

    c) Unbiased and precise d) Precise

19. There are N beads in a jar, 40 of these beads are black. Ram takes at random a sample of 50 beads from the jar, 5 of the beads in his sample are black. Work out an estimate for the value of N.

    a) 400 b) 450

    c) 120 d) 200

20. A farmer wants to estimate the number of rabbits on his farm, on Sunday he catches 120 rabbits. He puts a tag on each rabbit and then runs away. On Monday he catches 70 rabbits out of 70 rabbits 15 of them have a tag. Work out the estimate for the total number of rabbits on the farm.

    a) 1000 b) 120

    c) 500 d) 560

Key to MCQ's

1.	(c)	2.	(a)	3.	(d)	4.	(b)	5.	(b)
6.	(a)	7.	(d)	8.	(a)	9.	(b)	10.	(a)
11.	(d)	12.	(a)	13.	(b)	14.	(a)	15.	(a)
16.	(a)	17	(b)	18	(a)	19.	(a)	20.	(d)

Let us do

1. In which situations Non-Response Technique is used? Explain with the help of a suitable example.
2. Discuss in detail the Randomized Response Technique (RRT) when we study or estimate the sensitive characteristics of a population.
3. Explain the qualitative response model with the help of a suitable example.
4. Explain the quantitative response model with the help of a suitable example.
5. What is capture recapture sampling? How it is used in agricultural sciences.
6. Differentiate between the qualitative response model and quantitative response model.